International Marketing and Management Research

Series Editor
Anshu Saxena Arora, School of Business and Public Administration,
University of the District of Columbia, Washington, DC, USA

International Marketing and Management Research (IMMR) Series provides a forum for academics and professionals to share the latest developments and advances in knowledge and practice of global business and international management. The series is a uniquely positioned contribution of interrelated research papers across all business disciplines including: marketing, international business, strategy, digital strategy, organizational behavior and cross-cultural management, international marketing, international finance, global value chains, global supply chain management, sustainable innovations, e-business and e-commerce, social media, new product design and innovation, and business economics. Six volumes have been published under the IMMR series. Each research paper published in this series goes through a double-blind peer review process. Through the series, we examine the impact of cross-cultural issues, characteristics, and challenges with regard to global and sustainable business innovations; institutional and regulatory factors on international marketing and management issues; and the effects of institutional changes on global businesses with regard to both traditional and digital worlds. IMMR series fosters the exchange of ideas on a range of important international subjects, provides stimulus for research, and strengthens the development of international perspectives. The international perspective is further enhanced by the geographical spread of the contributors.

Anshu Saxena Arora · Sabine Jentjens ·
Amit Arora · John R. McIntyre ·
Mohamad Sepehri
Editors

Managing Social Robotics and Socio-cultural Business Norms

Parallel Worlds of Emerging AI and Human Virtues

palgrave
macmillan

Editors
Anshu Saxena Arora
University of the District of Columbia
Washington, DC, USA

Amit Arora
University of the District of Columbia
Washington, DC, USA

Mohamad Sepehri
University of the District of Columbia
Washington, DC, USA

Sabine Jentjens
ISC Paris Business School
Paris, France

John R. McIntyre
Scheller College of Business
Georgia Institute of Technology
Atlanta, GA, USA

ISSN 2662-8546   ISSN 2662-8554   (electronic)
International Marketing and Management Research
ISBN 978-3-031-04866-1   ISBN 978-3-031-04867-8   (eBook)
https://doi.org/10.1007/978-3-031-04867-8

Cover credit: © Melisa Hasan

This Palgrave Macmillan imprint is published by the registered company Springer Nature Switzerland AG
The registered company address is: Gewerbestrasse 11, 6330 Cham, Switzerland

Acknowledgments

Research involving Social Robotics, Artificial Intelligence, and Digital Technology (Sections I and II) is funded by the National Science Foundation Grants #1912070 and #2100934.

CONTENTS

EDITORS AND CONTRIBUTORS

About the Editors

Arora Anshu Saxena, Ph.D., PMP is Associate Professor of Marketing at the School of Business and Public Administration, University of the District of Columbia (UDC), Washington, DC, USA.

Jentjens Sabine, Ph.D. is Professor of Management at ISC Paris in France.

Arora Amit, Ph.D. is Associate Professor of Supply Chain Management in the School of Business and Public Administration at the University of the District of Columbia (UDC), Washington, DC, USA.

McIntyre John R., Ph.D. is Founding Executive Director of the Georgia Tech CIBER, a national center of excellence, emphasizing the role of techinnovation and sustainable systems in international business-related fields. He is Professor of Management (Strategy & Innovation Area) in the Scheller College of Business with cross appointment in the Sam Nunn School of International Affairs, Georgia Institute of Technology, Atlanta, Georgia. He received his Ph.D. from The University of Georgia at Athens, following graduate studies at McGill, Northeastern, and Strasbourg Universities. He is the author and co-editor of fifteen books, the latest: *The New Chinese Dream: Industrial Policy in the Post-pandemic Era*

(Springer, 2021) and forthcoming: *The Role of Multinational Enterprises in Supporting the UN SDG's* (E Elgar). He is the recipient of the State of Georgia's Governor's International Award for international business education excellence; the university-wide Steven Denning Faculty Award for Global Engagement; and the French National Order of Merit for work in promoting US-European FDI. He is currently an honorary professor at ICN Graduate Business School, University of Lorraine, near the Georgia Tech European Campus in Metz.

Sepehri Mohamad, Ph.D. is Dean of the School of Business and Public Administration at the University of the District of Columbia. Prior to joining UDC in 2015, Dr. Sepehri was the Associate Dean & Director of the Graduate Business Programs and the Coordinator of the Adult/Accelerated Degree Program at the Davis College of Business in Jacksonville University. Dean Sepehri has Triple-major Ph.D. in Management, Higher Education Administration, and Strategic International & Political Economy from Indiana University, Bloomington. Dr. Sepehri has over 35 years of experience in strategic planning and global competitive market analysis. He is a consultant and research analyst to business corporations, and a number of city, county, and state governments.

Contributors

Akpan Ndifreke holds a Master of Business Administration degree from the School of Business and Public Administration at the University of the District of Columbia (UDC), Washington, DC, USA.

Ampe-Nda Loucace Dorcas holds a Master of Business Administration degree from the School of Business and Public Administration at the University of the District of Columbia (UDC), Washington, DC, USA.

Arapiraca Carla Jamille holds a Master of Business Administration degree from the School of Business and Public Administration at the University of the District of Columbia (UDC), Washington, DC, USA.

Bacouel Anne-Sophie is a Ph.D. student at the University of St. Gallen in Switzerland. She graduated in International Business Administration from WHU Otto Beisheim School of Management in Germany (B.A.) and with a CEMS Master's in International Management from Rotterdam School of Management and Vienna University of Economics and Business.

Bacouel Victoria is a student in International Business Administration at Rotterdam School of Management in The Netherlands.

Bernhard Fabian, Ph.D. is Associate Professor of Management and part of the Family Business Center at EDHEC Business School in Paris, Lille, Nice, and London. He is Research Fellow for family business at the University of Mannheim and for psychology at the University of Frankfurt in Germany.

Buzolin Renata holds a Master of Business Administration degree from the School of Business and Public Administration at the University of the District of Columbia (UDC), Washington, DC, USA.

Hoek Ashley is a Dutch-American who holds a master's degree in business management from EDHEC Business School and a project management certification. Professionally, she works in the international development and nonprofit sector.

Kim Jaehwan holds a Master of Business Administration degree from the School of Business and Public Administration at the University of the District of Columbia (UDC), Washington, DC, USA.

Ohio Curley holds a Master of Business Administration degree from the School of Business and Public Administration at the University of the District of Columbia (UDC), Washington, DC, USA.

Osbourne Emily is Undergraduate Bachelor of Business Administration student at the School of Business and Public Administration, University of the District of Columbia (UDC) with a dual concentration in Accounting and Logistics & International Trade Analytics.

Patanachaisiri Prim holds a Master of Business Administration degree from the School of Business and Public Administration at the University of the District of Columbia (UDC), Washington, DC, USA.

Payne Barbara Ann holds a Master of Business Administration degree from the School of Business and Public Administration at the University of the District of Columbia (UDC), Washington, DC, USA.

Sammonds Andrew is Undergraduate Bachelor of Business Administration student at the School of Business and Public Administration, University of the District of Columbia (UDC) with a dual concentration in Finance and Logistics & International Trade Analytics.

Yepez Grace is Undergraduate Bachelor of Business Administration student at the School of Business and Public Administration, University of the District of Columbia (UDC) with a dual concentration in Marketing and Logistics & International Trade Analytics.

Associate Editor

Anyu Julius Ndumbe, Ph.D. is a Tenured Professor of Public Administration and Public Policy at the School of Business and Public Administration at the University of the District of Columbia, Washington, DC. He holds a Ph.D. degree in Political Science from Howard University. From August 2014 to April 2015, Professor Anyu served on four US State Department Boards that reviewed some personnel files of Senior Foreign Service Officers for performance pay and/or promotion. Professor Anyu has authored several well-cited scholarly peer reviewed articles. His most recent book is entitled: *The Lure of Chinese Money in Africa* (2022), His other published books include: *Case Studies of Conflict in Africa* (2013). *The Foreign Corrupt Practices Act: A Catalyst for Global Corruption Reforms* (2007), *Democracy, Diamond and Oil: Politics in Africa Today* (2006) and *People on the Move* (2004). Dr. Anyu currently serves assistant editor at the *International Journal of Organizational Innovation*, published by University of Indiana Press. In summer 2021, Professor served as Guest Editor at the Sustainability Journal. He previously served as assistant editor and book review editor at the *Mediterranean Quarterly*, (2000–2013) published by Duke University Press.

Editorial Review Board

LIST OF FIGURES

LIST OF TABLES

LIST OF APPENDICES

CHAPTER 1

Introduction

John R. McIntyre, Anshu Saxena Arora, and Amit Arora

The service sector has always been a laboratory for innovation. Technologies such as AI, clouding, and data banks have been implemented to revolutionize the future of society and industry. Whenever a new technology emerges, it induces fear and lack of trust. This volume, focused as it is on social robotics business norms and parallel worlds, is right at the core of the central question of trust and ethics in artificial intelligence. Some existing regulations could be applied to the notion of safety with robots but new ones will need inventing (Gualtieri et al., 2022; Wirtz, et al., 2018). According to the Association for Computing Machinery (ACM), professionals constructing and implementing such systems must attempt to ensure that the products of their efforts will be used in socially responsible ways, will meet social needs, and will avoid harmful effects to health and welfare (Klüber & Onnasch, 2022; Nagenborg, et al.,

J. R. McIntyre (✉)
Georgia Institute of Technology, Atlanta, GA, USA

A. S. Arora (✉) · A. Arora
University of the District of Columbia, Washington, DC, USA
e-mail: anshu.arora@udc.edu

A. S. Arora et al. (eds.), *Managing Social Robotics and Socio-cultural Business Norms*, International Marketing and Management Research, https://doi.org/10.1007/978-3-031-04867-8_1

1

2008). Today's deep learning systems which imitate neural networks in the human brain and can absorb vast amounts of data are able to teach themselves to perform some tasks, from pattern recognition to translation, almost as well as humans can. As a result, tasks which once called for a human mind are now within the scope of computer programs. A noteworthy example of such is the research work done at the Georgia Institute of Technology where Professor Mark Riedl has employed techniques to encourage a game-playing AI to explain its own moves (Georgia Tech School of Interactive Computing, 2022). The Georgia Tech team trained an AI agent to match these narratives to the internal features of second agent that had already learned to play 'Frogger.' The experiment produced a system offering snippets of human language that described the way the second agent is playing the game. It is this type of experiments that open the black box of AI, raising essential questions about the limits and the uses.

Ethical considerations which are reviewed in depth in this volume are at the heart of the challenges faced by social robotics and its broader artificial intelligence applications in various societal fields. Today, artificial intelligence (AI) is a protean phenomenon which can surpass humans in performance in many areas of human activity. However, AI reaches its limits in unfamiliar situations, i.e., that are not represented in the training data. In such situations, humans remain superior to AI because humans can draw on prior and general problem-solving knowledge and consider the terms of the problem in the relevant context. Researchers then draw on their experience as well as knowledge passed down to them over generations. If unable to solve a complex problem, the iterative process of science helps: Hypotheses can be formulated, experiments conducted as well as investigations, thus confirming or refuting assumptions. AI, however, does not have (or has limited), at this point, corresponding knowledge transfer/dissemination to the same depth as humans do. AI is gaining importance in the new world of business, while in the parallel world, humans (still) lead the way. Machines with artificial intelligence (especially, social robotics that interact with humans) analyze big industry data sets that humans cannot and utilize them to form models that are effective in both worlds. Touchless AI, robotic, and digital technology have emerged as a critical support systems for educational, therapeutic, and emotional needs during the COVID-19 pandemic. The COVID-19 global pandemic has changed how the new world is perceived where touchless robotic/digital technology (e.g., social-collaborative robots) emerged as a critical human-support business system for educational, automation, digitization, therapeutic, and emotional needs.

In this new era of diffused humanized technology and new world of the current COVID-19 pandemic, social robots are becoming ubiquitous. Nevertheless, we are still a great distance from AI and social robotics dominating key areas of life. Many challenges persist, including a true understanding of abstract concepts, transfer of knowledge to novel applications, transparency and security guarantees, distinguishing between random and logically meaningful relationships. While autonomous machines are valuable tools, only humans remain capable of recognizing values supporting ethical and socio-cultural norms. Human virtues, such as emotional intelligence, wisdom, and courage, are required for decision-making in many (private and professional) situations where machines would lead to sub-optimal and/or ethically questionable outcomes. Furthermore, humans can adapt behavior relatively quickly to new contexts, influenced by socio-cultural norms, leading to business negotiations nuanced in transitional (2-world) business situations. Knowledge or cognition includes the factual knowledge of other cultures, for example, the history of another country or the particular preferences of its population. This can of course be stored in detail via AI. However, individuals draw on the cultural imprint that they have experienced in their own cultural environment, leading to subtle knowledge which might not be captured or coded yet by AI.

While it is true that AI can learn, this learning is quite different from human learning. Humans can learn implicitly, that is, through play and experimentation. AI cannot learn implicitly, but only explicitly based on the data provided to it. The Riedl experiment noted above is an illustration of this phenomenon. One can argue that intelligent systems can improve their abilities by means of appropriate algorithms; however, if you make a small change to the rules, humans can readily adapt their individual and collective behavior. AI, on the other hand, is confronted with a completely new problem set and has to start from scratch and relearn with the rule changes. Such changes to the rules constantly happen in human interactions, where socio-cultural norms are at play. While AI has already made its way into the intercultural domain, e.g., in the translation of foreign languages, intercultural interaction has so far been AI-free and dominated by social and thus human values. While this compilation praises the benefits of AI and digital technology for areas of social life in its first and second parts, it is of interest to highlight those areas, where humans are not yet replaceable (in the last section).

In this volume, the authors present three sections. **Section I—Social Robotics**—encompasses chapters capturing the complex world of social robotics and AI. Chapters 1–4 (under Section I) focus on the emerging world of AI and Social Robotics. **Section II—Digital Technology**—addresses the complex details of digital technology, AI, and social media roles in business applications. Chapters 5–7 (under Section II) highlight digital media and technology and its impacts during the current turbulent times. Finally, in **Section III—Parallel Worlds: Human-Focused Socio-Cultural Norms and Negotiations—an AI-Free Research Domain**, the authors highlight human actions embedded within socio-cultural business norms (e.g., international negotiations, work culture, and religious diversity) that are not yet replaceable by AI. Chapters 8–10 (under Section III) highlight the parallel world of human values and socio-cultural norms that is AI free and human-focused.

Section I focuses on Social Robotics. Social robotics emphasizes on the social interaction between robots and humans. The field of social robotics deals with the evolution of robots that display emotional and social behavior, along with the examination of imitation behavior of robots, both socially and emotionally. Social robots display anthropomorphic/humanlike characteristics like mimicking human emotions/expression of emotions, facial/voice recognition, ability to conduct high-level dialogue, develop social competencies, ability to learn/develop personality, using natural cues, etc. **Chapter 1—From Robots to Humanoids: Examining an Ethical View of Social Robotics**—introduces a new concept called ethical dilemma of artificial intelligence (ED-AI), illustrating the value of sixth sense in social robotics. **Chapter 2—Advancement of Robotic Autonomy Benefitting Individuals with Autism: Ethical Curriculum Development through Social Robotics' Design and Research**—investigates designing and developing ethical curriculum through robotic autonomy, social cognition, ethics, and human engagement aimed at individuals with autism spectrum disorder (ASD). **Chapter 3—Recalibrating Anthropomorphized Robotic Interactions during COVID-19: Understanding Human Robotic Interactions**—focuses on anthropomorphism in social robotics, human personality dimensions, and emotional-behavioral changes related to robotic likeability and acceptance of overall human–robot interaction (HRI). **Chapter 4—Robotic Anthropomorphism and Intentionality through**

Human–Robot Interaction (HRI): Autism and the Human Experience—investigates social-cognitive intelligence in relation to artificial intelligence (AI) and robotic anthropomorphism while designing ethical, artificially intelligent, social robots for an overall successful HRI experience.

Section II focuses on Digital Technology. Digital technology/digital media can be defined as all electronic tools, automatic systems, technological devices, and resources that generate, process, or store information. Examples of digital technology are social software (e.g., wikis, blogs, social media, and social networks) and electronic communication capabilities (e.g., Web conferencing platforms like Microsoft Teams, Cisco WebEx, and Zoom) that are targeted to enable social interactions. This section is deliberately delineated from the previous section since this section focuses on digital technology and media with a strong focus on consumers during the current, turbulent times. **Chapter 5—** Artificial Intelligence (AI) in Marketing: How AI Supports Marketers throughout the Consumer Journey—examines AI's building blocks (e.g., natural language processing, image recognition, speech recognition, problem-solving and reasoning, and machine learning) in marketing for understanding overall consumers' needs and consumer decision-making process. **Chapter 6—**Digital Technology Roles for COVID-19 Crisis Management: Lessons from the Emerging Countries—focuses on the spread of COVID-19 global pandemic, and how digital technologies can help in aiding and improving every stage of COVID-19—before contracting the disease, during the disease, and after-treatment of the disease. **Chapter 7—**Social Commerce: Impact on Consumer Power through Social Media—examines consumer behavior and consumer power (or decision-making process) in both physical and virtual worlds, and the subsequent emergence of social commerce in the computer mediate social environments (CMSEs).

Section III focuses on Parallel Worlds: Human-Focused Socio-Cultural Norms and Negotiations. This human-focused section is deliberately kept as an 'AI-free domain' section with an emphasis on culture and cultural differences, women career advancement, and religious diversity. Even though, one may argue that AI is present in all spheres of life, yet we have maintained 'Section III' as a 'technology free' and 'human-focused' in our humble effort of highlighting human values, social relationships, and societal/cultural prominence. That is why, this section is referred to as 'Parallel Worlds' where both technology and human interference exist with a human-touch dominance.

Chapter 8—Cross-Cultural Negotiations and the Impact of Culture in a Western-Asian Context—focuses on cross-cultural negotiations by analyzing and comparing the different negotiation experiences that European informants had in East Asia. The chapter presents the negotiation experiences of European professionals in diverse industries who actively took part in negotiations in East Asia with East Asian counterparts. **Chapter** 9—The Effects of Work Culture on Women Career Advancement: A Comparison between the Netherlands and the USA—investigates seven distinct cultural workplace differences between the USA and the Netherlands and discusses practical implications. This chapter analyzes the effects on women and their career potential, and reveals how Netherlands can offer a more conducive workplace for women than the USA. **Chapter** 10—Respect or Transgression of Norms in the Context of Religious Diversity: The Example of the Mormons in France—examines how the controversy over the construction of the Mormon temple in France impacted the perception of Mormons as a potential religious outgroup in France. These three chapters build the foundation of 'cultural differences,' 'women career advancement,' and 'religious diversity': the human-centered concepts that are not be replaced by AI.

Dan Sperber, a cognitive psychologist at the Jean Nicod Institute in Paris, notes that autonomous machines and humans differ widely in constructing reasons for behavior. Humans align such reasons with their own interests; AI systems do not have their own interests to serve. Their explanatory frameworks are built by and for human beings. At what point shall we need to ascribe ethical, societal, and cultural personhood to AI systems? This volume provides deep insights in the rapidly growing expert literature.

References

Georgia Tech School of Interactive Computing (2022), https://ic.gatech.edu/news/586346/ai-translates-its-internal-monologue-humans-understand-and-plays-frogger

Gualtieri, L., Rauch, E., & Vidoni, R. (2022). Development and validation of guidelines for safety in human-robot collaborative assembly systems. *Computers & Industrial Engineering, 163,* 107801.

Klüber, K., & Onnasch, L. (2022). Appearance is not everything-Preferred feature combinations for care robots. *Computers in Human Behavior, 128,* 107128.

Nagenborg, M., Capurro, R., Weber, J., & Pingel, C. (2008). Ethical regulations on robotics in Europe. *AI & Society, 22*(3), 349–366.

Sperber, Dan, http://www.institutnicod.org/recherche/projets/?lang=fr

Wirtz, J., Patterson, P. G., Kunz, W. H., Gruber, T., Lu, V. N., Paluch, S., & Martins, A. (2018). Brave new world: Service robots in the frontline. *Journal of Service Management, 29*(5), 907–931.

PART I

Social Robotics

CHAPTER 2

From Robots to Humanoids: Examining an Ethical View of Social Robotics

Loucace Dorcas Ampe-Nda, Barbara Ann Payne,
Anshu Saxena Arora, and Amit Arora

Abstract Technology enables robots to have a voice, interact with humans, and execute a series of commands through artificial intelligence (AI) and human–robot interaction (HRI). The notion that robots can surpass humans implies that they can become conscious through five human senses along with ethics as the sixth sense. The sixth sense of ethics in robots can produce an efficient and versatile humanoid. Therefore, the

L. D. Ampe-Nda · B. A. Payne · A. S. Arora (✉) · A. Arora
University of the District of Columbia, Washington, DC, USA
e-mail: anshu.arora@udc.edu

L. D. Ampe-Nda
e-mail: loucace.ampe@udc.edu

B. A. Payne
e-mail: barbara.payne@udc.edu

A. Arora
e-mail: amit.arora@udc.edu

A. S. Arora et al. (eds.), *Managing Social Robotics and Socio-cultural Business Norms*, International Marketing and Management Research, https://doi.org/10.1007/978-3-031-04867-8_2

key concept that emerges in our chapter is the ethical dilemma of artificial intelligence (ED-AI) and addresses the question whether the sixth sense can play a role in the robotic paradigms and perform relationship-building behaviors with a positive interaction.

Keywords Artificial intelligence · Consciousness · Human–robot interaction · Robots · Robotic systems · Ethical dilemma of artificial intelligence (ED-AI) · Technology governance · Sixth sense

INTRODUCTION

John McCarthy, an American scientist in 1950s, is considered a co-founder of the field of Artificial Intelligence (AI) (Computerhistory.org, 2016). Since then, AI has come a long way and for many years scientists have been developing robots that could behave like humans. The twenty-first century seems to be fulfilling the desire of the previous generations to make our world more futuristic. For many years, scientists and philosophers have been debating about the nature of brain and its relationship with the mind. This leads to some fundamental issues including claims about new futurist robots that do not care about the importance of what it is to be a human, the necessities to build conscious machines, or its implications (Boada et al., 2021; Signorelli, 2018). Smart speakers, such as Amazon's Alexa, a virtual personal assistant, which is a wireless device with artificial intelligence can be activated through voice command. Through such AI algorithms, senses of these devices can measure up to some attributes when information is processed. AI enabled voice assistants can service an individual based on commands and accomplishments of input and output experiences. Some virtual assistants can understand human speech and respond via manufactured voices . Consumers can ask their virtual assistants to manage calendars, to-do-list, and emails.

Marx's dialectical theory propagates that every phenomenon must have two sides. When enjoying a positive influence, one cannot ignore its negative influence, especially in the development of science and technology (Ma et al., 2018). Since Marx's conception theories are to be understood as representations of socially regulated experience, theoretical critique here echoes Hegel's remark that dialectical consciousness is not "peculiarly confined to the philosopher," so that it "would be truer to say that dialectic gives expression to a law which is felt in all other grades of consciousness, and in general experience" (Hegel, 1892, pp. 149–150). Computers with their vital components such as AI have evolved into super-machines. One of the essential principles of Marx's dialectical theory shows the negative and positive impacts caused by all elements. In artificial intelligence, people enjoy positive interactions and at the same time, face the potential of taking over human intelligence. In voice, the merits of the theory include flexible interfaces for communication that are challenged by the ease of cyber-attack. The ethics field allows the anthropomorphisms of robots, but it is set back by the dilemma of giving our inventions human identities. This paper is an exploration of the Ethical Dilemmas of Artificial Intelligence (ED-AI). Modern legislation is not yet prepared to face the reality of artificial intelligence developments, and there is no guarantee that people will be fully secure in the brand-new world. There is a dearth of research in the field of social robotics, voice assistants, robot ethics, and sixth sense (Ma et al., 2018; Signorelli, 2018; Smith, 2018; Harrison, 2018; Boada et al., 2021). To bridge the gap, this research has the following objectives.

- Examine a framework that would allow conscious machines to meet and surpass human intelligence and capabilities;
- Investigate the kinds of AI interactions that feel more natural and lead us in our progression from chatbots to avatars to humanoid robots with voice, gaze, sixth senses, and other anthropomorphic qualities; and
- Examine how the law and a set of HRI regulations develop ethics to protect humans and impute liability for the limitations and failure of artificial intelligence.

This research consists of four sections. First, we focus on defining and describing artificial intelligence and how AI interacts with humans through voice assistants, displaying consciousness using sixth sense. Second, we examine how AI has impacted and continues to influence consumer behavior through robot ethics, consciousness, and anthropomorphism. Next, we utilize the Ethical Dilemmas of AI (ED-AI) framework to explore the benefits and risks associated with using the ethical approach in AI and social robotics. While there are always risks associated with social-ethical robotics, ED-AI has become an essential key to success today. Finally, we discuss the positive influences that ED-AI (through the utilization of social robotics) framework will have on business-to-business relationships, as well as business-to-consumer relationships.

Theoretical Background

AI still needs a considerable leap to reach the level of human intelligence. Aspects, such as autonomy, reproduction, and morality, make modern machines different from humans (Signorelli, 2018). It is necessary to understand how the human brain works and what our moral boundaries are to compare humans to AI. On the other hand, some researchers argue that AI is capable of all human activities, even those involved in creativity and interaction. Smith (2018) describes the world of Alexa robot, like a real human that can perform tasks people are not able to do. Machines take the shortest route between data input and output (Ma et al., 2018; Coeckelbergh, 2022). The human brain tends to follow a sequence of steps before arriving at the desired outcome.

AI has some limitations and deficiencies especially in the context of situations involving ethical dilemmas where human lives are endangered. In such instances, humans are replaced by machines to achieve efficiency (Ma et al., 2018). New technology has discovered that ethical (sixth sense) robots are a revolutionary way to engage and interact with humans. According to Horningold (2019) and Coeckelbergh (2022), new technology often involves new arrays of sensitive material and machine learning algorithms to fill in the gaps. However, cloning this sophisticated sense can pose a challenge for robots. Researchers at a Swiss University created a bionic hand that gives amputees a "close to nature touch" that reproduces the feeling of proprioception (Nichols, 2019). The device stimulates nerves that provide sensory feedback in real-time. Though the challenge was to have similarities with human sense, the development of using the model is observed to provide a solution through the senses. Robots are machines capable of sense, plan, and act with human interaction. Robots need sensors to mimic social abilities and enhance their task. When robots have sensors, this allows them to reach and explore beyond human capabilities.

ARTIFICIAL INTELLIGENCE AND SOCIAL ROBOTICS

Marx's dialectical theory is one of the critical theories that is applicable and relevant in the domain of social robotics. In essence, this theory is concerned with the multifaceted nature of the phenomenon in all life aspects and the need to consider all these when engaging with fellow human beings and the more recent trend of artificial intelligence. Specifically, in relation to artificial intelligence, there are various inherent positive and negative facets, and it is essential to consider both dimensions to develop a balance (Walker, 2012). While there may be the benefit of better analytics and augmented functions, there is also the ethical conundrum of application.

SOCIAL ROBOTICS AND VOICE OR SIXTH SENSE

Humans have an innate tendency to anthropomorphize surrounding entities and have always been fascinated by the creation of machines endowed with human capabilities and traits (Hudson & Mankoff, 2014). The example of the hierarchy paradigm below consists of gathering and processing data received through a robot's sixth sense sensors. The processed data are used to demonstrate the human and robot interaction. The model established a set of symbols collected by predictions that can be operated by a logical system with humans to robotic machines. Once a suitable set of actions has been made, this helps executes the effect by changing a high-level command (voice) to mid-level (sense) into low-level commands (act) using a robot cognitive system. The process is repeated continuously until the primary goal of human interaction and AI has been achieved. The goal is to create robots that can interact with humans and move around in unity while understanding the series of voice, sense, and act.

Further, the testing of human–robot interaction through social behavior has been analyzed to be natural and familiar. The "voice, sense, and act" (see Fig. 2.1) provide a communication channel through which action can coordinate and control any information distributed from humans to robots. According to researchers at Nanyang Technological Institution in Singapore, after interpreting the voice command (voice), a series of control data for performing tasks are generated (sense), and the robot finally performs a task (act) (sciencedirect.com). In the later sections, an AI machine named "Alexa" is described that can function

Fig. 2.1 Hierarchy Paradigm based on Voice, Sense, and Act

and communicate with humans through voice, sense, and act processing systems.

SOCIAL ROBOTICS AND ANTHROPOMORPHISM: EXPLORING ROBOTIC CONSCIOUSNESS

As AI gradually bridges the gap with human intelligence, it is necessary to explore significant aspects that raise ethical dilemmas associated with it Robotic consciousness is the ability of robotic machines to interpret and to analyze data similarly to human beings. Since humans develop artificially intelligent systems, this essentially translates to an ultimate situation whereby robotic intelligence fully surpasses that of humans. This characteristic is further established by the recent unconscious wave of the anthropomorphism of robots.

In past cases, artificially intelligent robots that were programmed with languages initially began to communicate, but it was sometimes impossible to decode by human developers. The study had to be stopped immediately with questions arising on the limitless competency of robots (Kiggins, 2018). Since limitations, such as human control and reliance on human intelligence, are currently dictating the course of robotics, robots are still under the control of humans. However, equipped with so many limitless capabilities, it is a case in point that it might take only a few years before robots completely surpass human capabilities. The consciousness of robots remains a dilemma in the technology field due to the frequent upgrades of these systems to make them as close to human as possible. Strong illustrations include Sophia, a robot equipped with machine learning program (Knox el at., 2019; Belanche et al., 2021; Coeckelbergh, 2022). Sophia was made the first robot citizen of Saudi Arabia, giving her probably more rights than local women. Along with her, Bina48, Philip, and Han are famous humanoids from Hanson Robotics located in Hong Kong. They are preprogrammed to hold a conversation using machine learning, natural language processing, and animated robotic software, even if they don't fully understand the meaning of everything. These inventions are almost better than the average human teacher. Along with characteristics of competence and diligence, these systems might gradually be absorbed, hence, negating the need for human tutors completely.

Conceptual Framework

Advancements in science and technology have led to the use of artificial intelligence that has impacted human lives. Innovations such as smart speakers and smart robots use artificial intelligence to conduct their duties. Ethical dilemmas surrounding the use of artificial intelligence revolve around consciousness, ethics, voice, and sixth sense.

There are concerns as to whether positive moral thinking can be achieved by improving learning algorithms or computational capability. These improvements may not necessarily make a machine conscious (Signorelli, 2018). Attempts to develop sentient machines are a wrong approach, as is trying to make them surpass humans. According to Signorelli (2018), futurists talk of super-machines that exhibit human characteristics without pausing to think about what it entails to be human. It is preposterous for science to emphasize anthropogenic presumptions or compare the intelligence of humans and robots.

Questions may arise on who will bear the responsibility for damages caused by the robots. A self-driving car and a surgical robot may cause a road or a medical accident, respectively. The law may regulate these incidents after they have occurred, but the damage will have been done. The deficiencies in the use of artificial intelligence may put the lives of people at risk. The use of smart speakers such as Alexa may expose the financial and personal information of people to hackers.

The voice of a social robot or a personal assistant in a smart speaker may be hard for listeners to understand. It may happen to persons who come from non-English speaking countries. There is also a chance that some listeners may feel offended by the accents or dialects adopted by the personal assistant on smart speakers.

Ethical Dilemmas of Artificial Intelligence (ED-AI)

The Ethical Dilemmas of Artificial Intelligence (ED-AI) framework (see Fig. 2.2) supports Marx's dialectical theory. Robots are vital elements of the scientific and technological fields that have proven their various irreplaceable positive influences (Walker, 2012). These include the consciousness of robots and the ease of anthropomorphism due to their similarity with real humans, such as voice and physical attributes. The dilemmas are levied by the limitless power of robots that can be consciously harmful in ethical terms. This observation means that the disadvantages might eventually outweigh the benefits.

The idea of Artificial Intelligence (AI) has been developed to a miscellany of controversial viewpoints regarding the current challenges of Human-Robots Interaction. A significant quantity of successful projects, already allowing the implementation of smart applications to optimize routine work, has proven that despite a range of capabilities machines afford, the measurement of their interference into real life is a complex riddle on the crossroads of scientific fields. Social interaction with robots should be discussed in the perspective of ethical challenges. For example, a self-driving car possessing consciousness is programmed to optimize the route in certain traffic conditions, although it could be challenging to measure the legal responsibility of a machine in case of a traffic accident. The human factor of immediate decision-making enables people to mentor robots, not vice versa. A sensor of a smart robot, which possesses certain anthropomorphic qualities, cannot be treated the way human senses are. Thus, artificial intelligence remains an object of direct human control. On the contrary, the evaluation of actual capabilities of the brain of robots enables people to optimize and automatize routine duties to such extent, that several jobs will be transferred to machines only. Curious management issues may be proposed here for negotiation: How is it possible to teach a machine the difference between human unemployment and robot leisure? Perhaps, the machine learning to generate human preferences through cases (algorithms for robot understanding) may be experimentally practiced as a precondition to effective mutual understanding. Therefore, a direct co-operation, and, consequently, communication with subjects of artificial intelligence have already become the crucial challenge people have ever had in management.

There is no doubt the numerous advantages of the successful implementation of smart technologies should not be underestimated. Nevertheless, the exploitation of artificial intelligence will lead to pending ethical problems regarding the future of human–robot coexistence: partial unemployment, inequality, behavioral dependency, insecurity, legal regulation, and original creativity (Bossmann, 2016). The management of AI

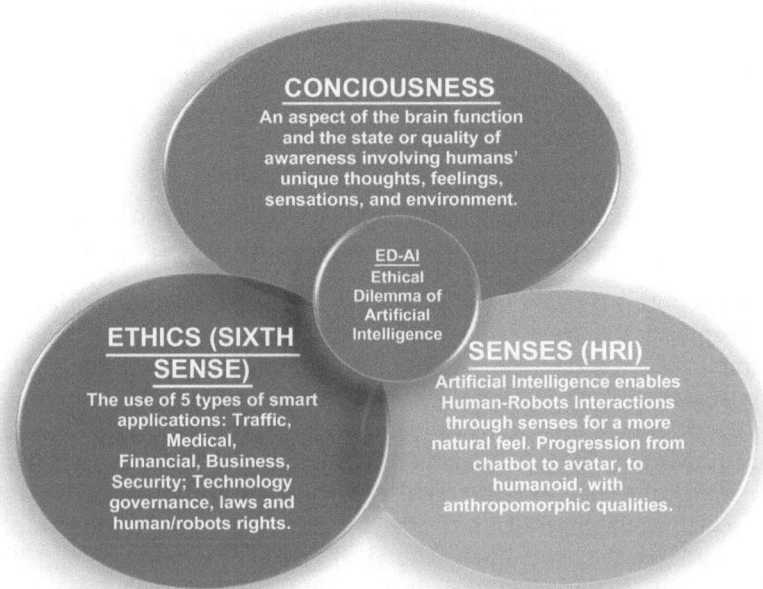

Fig. 2.2 ED-AI Framework in Social-Ethical Robotics exploring Sixth Sense: How "Robot Ethics" impacts "Robotic Consciousness / Anthropomorphism" and "Voice"

performance is currently a scantily investigated topic, which integrates a range of approaches and sciences. Therefore, an immediate need for a human–robot relationship solution remains the key point of modern ethical studies.

Voice Assistants, Consciousness, and ED-AI

There are various aspects of social robotics that need to be evaluated to achieve the envisioned objectives as enshrined by AI. In essence, a scenario where technological advancements, including social robots and smart speakers, can be programmed to adhere to ethical standards, they manifest consciousness that can enhance human–robot collaboration (Ma et al., 2018; Belanche et al., 2021). However, the concept of voice assistants is characterized by various ethical dilemmas that are inherently attached to their functionality. Thus, we offer the following proposition.

Proposition 1: *When technological advancements such as smart speakers and social robots follow ethical standards, they exhibit consciousness for better human–robot interaction.*

ED-AI: Relationship to Robotic Anthropomorphism and the Sixth Sense

Social robots are considered an entertainment relationship with anthropomorphism and have evolved into social interaction with humans. For example, supposedly, the paradigmatic shift emphasized the role and importance of the body and the environment in the cognitive competence of robotic agents, that leads to giving more considerable attention to the social environment as a fundamental factor in cognitive capabilities and development. Therefore, the goal is not merely to artificially reproduce the social intelligence of human agents but centralize personal information with social robots (Belanche et al., 2021; Damiano et al., 2015). Additionally, the idea is to actively include users in the social performances and presence of a robot by controlling robotics agents that inspire users to feature human feelings and mental states to robots, which should enhance and encourage social interactions. Thus, we offer the following proposition.

Proposition 2: *When machines/robots are considered as intelligent, humans anthropomorphize robots due to a robot's association to voice and consciousness.*

Ethics and ED-AI

Ethics are correlated with the ethical dilemma of AI in many ways. As AI systems move toward attaining human intelligence, discussions, and laws are recognized and pushed for the robots' rights like the citizenship of robots. While this is a feasible appreciation of robotic qualities, it is vital to understand the massive progress that the field continues attaining (Belanche et al., 2021; Kiggins, 2018). By allowing robots with this kind of power, a dilemma arises, whereby total robotic consciousness is a factor that can provoke human oppression. Thus, we offer the following proposition.

Proposition 3: When machines/robots are considered as intelligent, humans begin to recognize robot laws/rights and ethical standards for robots and other robotic systems.

Discussions

Companies worldwide need to understand and utilize social-ethical robotics so they can thrive in the new age and reach maximum consumers. Our research can guide this task by explaining AI, social robotics and ED-AI, and illustrating why companies will need new strategies and actions in this new ED-AI era, especially dealing with social robotics, sixth sense, and AI. Business managers can use this information to differentiate themselves from their competitors. Taking into consideration the fact that advanced machines already exist, it can be inferred that robots may develop a better communication process with humans over time. Machines' interactive abilities allow them to take on the position of a professional advisor. Humans develop trust only if an object is in proximity and capable of maintaining eye contact. Such robots are more efficient in targeting the advertisement and search for the best option when it comes to investments, shopping, or travel planning.

The interaction of the Ethical Dilemma of AI and Voice has positive impacts, such as better communication and the ease of user access; however, the negative impact is ease of hacking and data breach. Better communication means consumers can use the intelligent systems to maximum potential, while also dealing with data security problems. The lack of ethical standards is challenging for everyone who uses social robots such as Alexa at home or at their workplace. Users become accustomed to a machine for so long that they can suffer separation anxiety or psychological trauma when the perceived emotional support is no more. It is dangerous to move too soon or too late as for the implementation of robot ethical rules. Sometimes the laws can obstruct creativity in technology. On the other hand, pushing for laws at a late stage of technology development can make those laws outdated too soon since the world of technology progresses at a rapid rate.

Both consciousness and anthropomorphism relate to ethical dilemmas by presenting merits, including cognitive computing that are hindered by the insecurity that this feature brings. Although computers enhance processes by learning from inputs, this enlightens programmers to their ability to surpass human intelligence. Ethics explores the probability of

AI systems to execute smart applications in hospitals, traffic regulation, and business settings. An emerging issue is the exploration of the citizenship of robots, a factor that might negate the control of humans over robots. The US IEEE standard is a global initiative that mandates that all things robotics should have protocols regarding security and privacy issues (IEEE, Robotics and Automation Society, 2019).

Modern AI machines still are not on the same level as humans. For example, Tailor Brands can develop unique logo designs based on the customers' preferences for a low price, which is a perfect solution for low-cost startups and shows that AI really can perform creative tasks learning from itself and people, which is called cognitive computing. The creativity of the system is limited to algorithms that evaluate the customer's descriptions and choices. As a result, the outcome may not evolve into a successful and outstanding logo, which is the main goal of designers (Avery, 2018). There is also no guarantee that such assimilation of machines into our daily lives will be safe in terms of rights. For example, if the machine advisor makes a mistake, there may be no one responsible. The existing legislation does not presuppose such cases. Thus, the probable solution depends on the ethical dilemma of artificial intelligence, which considers whether machines are treated as humans, which is still an unresolved issue. The British Standard guide to the ethical design and application of robots and robotics systems allow roboticist and designers to perform an ethical risk assessment of these artificial agents (Villaronga, 2019).

Sophia is AI in its infancy stage, heading to artificial general intelligence (AGI) (full human intelligence), which has not yet been achieved. At this stage, robots will be able to communicate with each other while sharing data through the AI mind cloud. Hanson is raising Sophia like a safe and good AI child. Some are against it and call her a puppet and a deceiver. Others think she may be doing more harm than good. The ethics lines are blurred, so there is a push for responsible foresight in the AI industry. There is a rising wave of AI ethicists dedicated to seeing AI and technology develop responsibly. If the desire is to make robots in such a way that there is reciprocal love or a human connection, people might be setting themselves up for disappointment. Nevertheless, Hanson said if we do not humanize robots and make them part of our family, the future might be frightening.

AI with sixth sense will have a more significant impact in the following decades. Researchers in Poland have developed AI with a sixth sense:

giving the ability to tell the difference between iris scans from living and dead people with a 99 percent accuracy (Harrison, 2018). Such advancements can potentially be the new algorithm for futuristic tools. The common-sense rule is that AI will not fully operate without human interaction. AI should not be treated as a single system; instead, it should be integrated with different solutions and capabilities (Stancombe, 2018). There should be a plan to have all five senses working together if humans want to replicate human intelligence in AI machines, along with development of sixth sense in intelligence automation.

At this point, research has demonstrated that AI still could not surpass humans in many areas. Its implementation is limited to human assistance rather than performing independent tasks. AI still has a massive potential in HRI. The higher the measure of consciousness, the higher the value. Although it may be harmful as there is no legislation securing human rights with the involvement of machines, taking control of the work of AI will eliminate the margin of error, hence the implementation of technology governance. Otherwise, it may cause much harm to humans. Cognitive and mental simulations can enhance robot's resilience to errors/failures (Boada et al., 2021; Bongard et al., 2006; Coeckelbergh, 2022; Vanderelst & Winfield, 2018).

Conclusion

As the world advances toward full adoption of artificial intelligence in its various forms, including voice assistance, consciousness, ethical dilemma, and sixth sense, there is a need to consider all the factors surrounding the phenomenon. Artificial intelligence is a significant aspect of the development of computers due to the range of merit it has generated. In essence, this is based on Marx's dialectical theory, which seeks to establish the various dimensions of the positive and negative aspects of AI. Ethical standards are imperative in the making of conscious and anthropomorphic machines and in the interaction of humans and robots. We shed light on a new generation of ethical robots and their technological governance. The robotics industry and the laws that govern them have some ground to cover because there is no consensus on how to legislate and regulate the robotic social sphere. There are many reasons to be cautious, as well as be hopeful in the progression of AI development. Therefore, designers/programmers must take control to ensure that Artificial Intelligence works toward generating more benefits than provoking harm. This research can go further and has the potential of adding the conjunction of emerging sophisticated ethical machinery with transferable human to robot interaction. The desire to invent a sixth sense robot has been emerging quickly. The advancement of technology has engineered a milestone in the development of sixth sense technology. The advancement of sixth sense technology integrated with other intelligent systems will usher in a new era in human–robot interaction.

References

Avery, J. (2018, August). Tailor Brands: Artificial Intelligence-driven branding. *Harvard Business School* Case 519–017.

Belanche, D., Casaló, L. V., Schepers, J., & Flavián, C. (2021). Examining the effects of robots' physical appearance, warmth, and competence in front-line services: The Humanness-Value-Loyalty model. *Psychology & Marketing, 38*(12), 2357–2376.

Boada, J. P., Maestre, B. R., & Genís, C. T. (2021). The ethical issues of social assistive robotics: A critical literature review. *Technology in Society, 67*, 101726.

Bongard, J., Zykov, V., & Lipson, H. (2006). Resilient machines through continuous self-modeling. *Science, 314*(5802), 1118–1121.

Bossmann, J. (21 October. 2016) *Top 9 Ethical Issues in Artificial Intelligence.* World Economic Forum. https://www.weforum.org/agenda/2016/10/top-10-ethical-issues-in-artificial-intelligence/

Brockman, B. K., Rawlston, M. E., Jones, M. A., & Halstead, D. (2010). An exploratory model of interpersonal cohesiveness in new product development teams. *Journal of Product Innovation Management, 27*, 201–219.

Coeckelbergh, M. (2022). The Ubuntu Robot: Towards a Relational Conceptual Framework for Intercultural Robotics. *Science and Engineering Ethics, 28*(2), 1–15.

Computerhistory.org. (2016). John McCarthy Computer History Museum. [online] Available at: http://www.computerhistory.org/fellowawards/hall/bios/John,McCarthy/ [Accessed 15 August. 2016].

Damiano L., Dumouchel P., & Lehmann H. (2015). Human-robot affective co-evolution. Int.j.Soc. Robot.77–18. [Cross Ref]. [Google Scholar].

Glaser, B. G., & Strauss, A. L. (1965). Discovery of substantive theory: A basic strategy underlying qualitative research. *American Behavioral Scientist, 8*(6), 5–12.

Harrison, L. (2018). Research develop AI with a sixth sense. Retrieved: www.govtech.com

Hegel, G. W. F. (1892). The Logic of Hegel, the encyclopedia of the philosophical sciences. (W. Wallace, Trans.). Oxford University Press. https://doi.org/10.1037/12966-000

Hornigold, T. (2019). Sensors and Machine Learning are Giving Robots a Sixth Sense. Retrieved: www.singularityhub.com

Hudson, S. E., & Mankoff, J. (2014). Concepts, values, and methods for technical human–computer interaction research. In Ways of Knowing in HCI (pp. 69–93). Springer.

Kiggins, R. D. (2018). *The political economy of robots: Prospects for prosperity and peace in the automated 21st century.* Palgrave Macmillan.

Knox, J., Wang, Y., & Gallagher, M. (2019). *Artificial Intelligence and Inclusive Education: Speculative Futures and Emerging Practices.* Springer.

Mabert, V. A., Muth, J. F., & Schmenner, R. W. (1992). Collapsing new product development times: Six case studies. *Journal of Product Innovation Management, 9,* 200–212.

Nichols, Greg (2019). A robotic hand that gives sixth sense back to users. *ZDNet.com*

NY. IEEE Robotics and Automation Society. (2019). IEEE Transactions on Robotics.

Sciencedirect.com (2003). *School of Electrical and Electronic Engineering,* Nanyang Technological Institute. Retrieved: 8 July. 2021.

Signorelli, M. C. (2018). Can Computers become Conscious and overcome Humans? *Frontiers in Robotics and AI, 5,* 121.

Silverman, D. (2000). *Doing qualitative research.* Sage Publications.

Smith, K. T. (2018). Marketing via smart speakers: what should Alexa say? *Journal of Strategic Marketing,* 1–16.

Stancombe, C. (2018). Is Cybersecurity the sixth sense of AI? Retrieved: http://www.capgemini.com

Vanderelst, D., & Winfield, A. (2018). An architecture for ethical robots inspired by the simulation theory of cognition. *Cognitive Systems Research, 48,* 56–66.

Villaronga, E. F. (2019). Robots, standards and the law: Rivalries between private standards and public policymaking for robot governance. *Computer Law & Security Review, 35*(2), 129–144.

Walker, G. E. (2012). *Toward a dialectic of philosophy and organization.* Brill.

Weller, C. (2017). Meet the first-ever robot citizen—a humanoid named Sophia that once said it would 'destroy humans. *Business Insider,* 7–10.

CHAPTER 3

Advancement of Robotic Autonomy Benefiting Individuals with Autism: Ethical Curriculum Development Through Social Robotics' Design and Research

Grace Yepez, Anshu Saxena Arora, and Amit Arora

Abstract The increased ability of robots to make unsupervised decisions necessitates mechanisms to ensure the safety of their behavior. This research examines the efficacy of social robots' design and research through curriculum development aimed at individuals with autism spectrum disorder (ASD). The study highlights the impact of social robots as 'assistive tools' for improving achievement in learning and providing

G. Yepez · A. S. Arora (✉) · A. Arora
University of the District of Columbia, Washington, DC, USA
e-mail: anshu.arora@udc.edu

G. Yepez
e-mail: grace.yepez@udc.edu

A. Arora
e-mail: amit.arora@udc.edu

© The Author(s), under exclusive license to Springer Nature Switzerland AG 2022
A. S. Arora et al. (eds.), *Managing Social Robotics and Socio-cultural Business Norms*, International Marketing and Management Research, https://doi.org/10.1007/978-3-031-04867-8_3

27

enriched and engaged learning experiences to ASD individuals. Inspired by the simulation theory of cognition, the research attempts to address ethical-social robotics by adding an ethical layer to a humanoid NAO robot. We focus on developing a STEM-focused 'robotic artificial intelligence (AI)' social robotics course for ASD individuals encompassing social-behavioral and ethical elements of social cognition, ethics, and basic and advanced rules of human behavior and engagement in human–robot interaction scenarios. The research highlights curriculum design and instruction processes, followed by the educational analysis methodology and key findings, demonstrating the massive beneficial impact of social-ethical robotic autonomy on student performance that extends far beyond the boundaries of specific technical concepts taught in robotics and helps ASD individuals develop social-behavioral, ethical, and computational thinking skills.

Keywords Simulation Theory · Autism Spectrum Disorder (ASD) · Robotic Autonomy · Ethical Layer · Course Curriculum · Student Performance · Computational Thinking Skills · Human–Robot Interaction

Introduction

A developmental disability distinguished by persistent deficits in social interaction and the existence of constrained, repetitive patterns of behaviors, interests, and activities is known as autism spectrum disorder. Improving learning, communication, and interaction requires early diagnosis and intervention. In the last three decades, ASD rates have risen significantly. According to the CDC, 1 in 54 children in the United States is identified with autism spectrum disorder.[1]

Prior research indicates that children with ASD have a difficult time keeping up with their usually developing peers. As a result, if these children do not receive some form of intervention and an appropriate structure, their knowledge and skill development may suffer. Previously conducted research found that incorporating assistive technology (AT) is

---

[1] https://www.autismspeaks.org/autism-statistics-asd.

advantageous in a variety of ways. The supply of appropriate AT is critical for helping students with ASD, and developing an ethical layer is imperative.

Autonomous robots are intelligent machines able to execute tasks in the world without human control, and the interdisciplinary study of the interaction between humans and robots is referred to as human–robot interaction (HRI). Based on prior research, there is evidence that robots can be successfully used as assistive technology (AT) in many fields, including social sciences, communications, and education. However, there is still a dearth of studies on how 'social robots' can be utilized as 'AT tools' to help children with ASD learn. As a result, research studies provided evidence that AT is effective in special education and inclusive settings.

Since social robots have the capability to teach students rigorous mathematical and scientific reasoning, it is regarded as a critical part of Science, Technology, Engineering, and Mathematics (STEM) education. For instance, research demonstrating how AT facilitates difficult or demanding learning tasks that previously appeared unattainable for children with ASD to comprehend will be presented with a STEM-focused curriculum through social robots and how robotic design and curriculum development impacts ASD. Robotics can be used to study applied mathematical ideas, the scientific method of inquiry, and problem-solving (Bers, 2010). Robots can mimic human movements, allowing students to construct mental representations of abstract mathematical concepts (Kopcha et al., 2017; Han, 2013; Kennedy et al., 2014).

Additionally, when students are confronted with complicated learning settings, the usage of robots can improve student motivation and encourage determination. The mathematical and scientific thinking connected with robotics education is referred to as computational thinking, which is a process that involves using core computer science concepts to solve issues, build systems, and comprehend human behavior (Kopcha et al., 2017; Wing, 2006). Computational thinking encourages processes such as abstraction, problem analysis, prediction, repetitive, repetitive thinking, and problem detection, which are all important in mathematics and science behavior (Barr et al., 2011; Grover & Pea 2013; Sengupta et al., 2013).

The topic is driven by a lack of existing materials and a desire for a curriculum that directly targets specific STEM requirements in an integrated approach targeted at social robotics and ASD individuals with learning disabilities. We address the following research questions.

- What problem-solving scenarios and explorations allow our newly developed STEM curriculum to help students cultivate computational thinking skills?
- What specific ways can robots with an ethical STEM-based curriculum provide valuable and enriching learning experiences with ASD?
- Lastly, how does the simulation theory of cognition help incorporate an ethical STEM focused curriculum aimed at ASD students?

The main purpose of this research is to focus on ethical-social robotics through simulation theory of cognition in social robotics, and how curriculum related scenarios developed for ASD individuals can result in improved social-cognition for ASD individuals. This article consists of four sections. First, we focus on defining and describing robotic autonomy in social robotics. Second, we examine how social-cognition traits of ASD individuals can be improved through robotic autonomy and STEM focused curriculum development using social-ethical robots targeting ASD individuals. Thereafter, we investigate the effects of robotic autonomy on ASD individuals, and the subsequent effect of human–robot interaction (HRI) scenarios and social-ethical interrelationships on the success of HRI implementation. Next, we propose our robotic autonomy—Curriculum Development—Social-Cognition framework and propose managerial implications of our framework on consumers and businesses in the context of ethical-social robotics.

Theoretical Background

When robots have an increasing ability to make unsupervised decisions, this is known as robotic autonomy. As a result, mechanisms must be in place to ensure the safety of the robot's behavior. The importance of equipping robots with mechanisms to ensure safety is further highlighted by the fact that many robots are designed to interact with people (Vanderelst & Winfield 2018; Royakkers & van Est, 2015). The development of safe robots is necessary, but not enough. These smart autonomous robots should also be consistently ethical. These robots should be capable of making decisions with underlying precautionary measures to prevent harm to any human (Vanderelst & Winfield 2018; Anderson & Anderson, 2007; Moor, 2006).

Autism, or autism spectrum disorder (ASD) is one of the five types of Pervasive Developmental Disorders (PDDs) that needs 'cognitive rehabilitation' as an educational process aimed at reducing the learning/cognitive disability within the limitations imposed by available resources, according to the Academy of Neurologic Communication Disorders and Sciences (ANCDS) (Beeson & Robey, 2012), and Diagnostic and Statistical Manual of Mental Disorders: Fourth Edition (DSM-IV) (Diagnostic and statistical manual of mental disorders: DSM-IV-TR, 2000). An ongoing neurological condition that leads to communication, social interaction, and behavior deficiencies is autism spectrum disorder (ASD). The degree of ASD-related impairment ranges from serious to near-typical social operations across a wide range of things. For some, the quality of their lives can be greatly affected. This includes but is not limited to severely constrained mobility, self-care, and interaction with others. Young people with autism spectrum disorder (ASD) are less likely than their mainstream peers to enter the workforce after finishing school; this is because communication deficiencies are a major impediment to workforce participation. (Bradford, Roberts-Yates, Silvera-Tawil, 2018).

About half of the children with ASD have insufficient spoken language for communicating effectively, and many of them never develop functional speech. Some will express their needs through the use of non-verbal cues, whereas others will communicate in phrases or sentences that have little to no meaning to others. Interpersonal communication can be challenging due to the large percentage of children with ASD who suffer from social anxiety, a dearth of spontaneity, and have difficulty initiating verbal and non-verbal communication with others (Bradford, Roberts-Yates, Silvera-Tawil, 2018).

Researchers have been investigating the use of socially assistive robots (SARs) to augment traditional education for autistic children for over a decade. The development of skills such as facial expression recognition, shared attention, imitative free-form play, and turn-taking has been demonstrated through the use of autonomous socially assistive robots (SARs). The most effective approaches have been those that utilize robots in free or semi-structured interactions (Bradford, Roberts-Yates, Silvera-Tawil, 2018).

The primary goal of socially assistive robots (SARs) is the development of social and communication skills in the field of ASD intervention. The majority of the work, however, has concentrated on the development of non-verbal communication skills such as eye contact and turn-taking Bradford, Roberts-Yates, Silvera-Tawil, 2018; Boccanfuso, O'Kane, 2011) There is a dearth of proof on the long-term benefits of SARs on this demographic's verbal communication skills. Children with autism enjoy learning from computers and mechanical devices. To facilitate interpersonal communication popular technologies such as apps, computer games, and devices for virtual reality are used by children with autism spectrum disorder (ASD). This technology provides reliable and realistic scenarios which can be constructed for representing everyday social situations. This technology also helps provide an environment for smooth learning and immediate feedback while minimizing the need for social 'real world' interactions during a process of learning, which is a common area of tension for many persons with autism.

The **simulation theory of cognition** is the theory of mind that allows for the richest representations (Hesslow, 2012; Wilson, 2002; Vanderelst & Winfield, 2018). It proposes that thinking executes the same cognitive (and neural) processes as interaction with the outside world. When thinking, actions are covert and are assumed to produce sensory inputs that evoke further actions, through associative brain mechanisms (Hesslow, 2012; Vanderelst & Winfield, 2018). Thinking, in this view, necessitates the construction of a concrete representation of the environment and not one that is made up of abstract symbols. Rather, it is assumed that experiences are represented by concrete instances and recombined using the brain's perception, action, and emotion systems.

This method used artificial reasoning procedures to determine whether robotic conduct complied with a set of established ethical criteria. In the same way, as Good Old-Fashioned AI (GOFAI) depended primarily on abstract symbolic reasoning or robot ethics, our approach to ethical robots does not rely on the verification of logic claims. We rely on internal simulations, which allow the robot to simulate actions and forecast their outcomes. (Mackworth, 2011; Vanderelst & Winfield, 2018). As a result, our technique is a type of robotic imagery. Robotic imagery has previously been used in various other fields of robotics. In their analysis of robotic imagery, (Marques & Holland, 2009; Vanderelst & Winfield,

2018) developed the term functional imagination to describe the process through which robots discreetly simulate behaviors and their outcomes to guide their future behavior in their analysis of robotic imagery. We'll use their term here and promote functional imagination as an approach for ethical robots.

We want to put in place consequentialist ethics, which is inherent in many cultures' and traditions' conceptions of morality (Haines, 2015). As a result, creating an architecture that is appropriate for this type of ethics is a good place to start. Furthermore, the ability to test the outcome of potential behaviors (Hesslow, 2002, 2012) without committing to them is a major benefit of a functional imagination (Marques & Holland, 2009; Ziemke, Jirenhed, & Hesslow, 2005). As a result, functional imagination can be used to support consequentialist ethics.

Conceptual Framework

As shown in Fig. 3.1, a set of potential behavioral choices is generated by the robot controller (a). The robot controller transmits the set to the ethical layer (b) to be checked before performing one of these options (c). The Simulation Module is used to test each potential behavior (d). This module simulates the motor and sensory repercussions of each behavior in the set, as well as the resulting interior states of the human and robot, using the present state of the world, human, and robot as a starting point. The Simulation Module transmits the projected internal states of the robot and the human to the Evaluation Module for every behavioral possibility (e).

It is often presumed that autistic children can build a functional set of daily living skills that enhance their chances of a more independent life by using first-person, realistic, and computer-generated environments. The reality of computer-based environments and the increasing sense of presence provided by interactive virtual environments have been argued for helping promote learning and increasing the likelihood of a person generalizing newly acquired skills into daily living. This is where the incorporation of autonomous social-ethical robots should come into play. A robot can produce complex behavioral patterns such as interpersonal interactions and generate social behavior, social perception, and appear less intimidating and predictable than people. Thus we offer the following propositions.

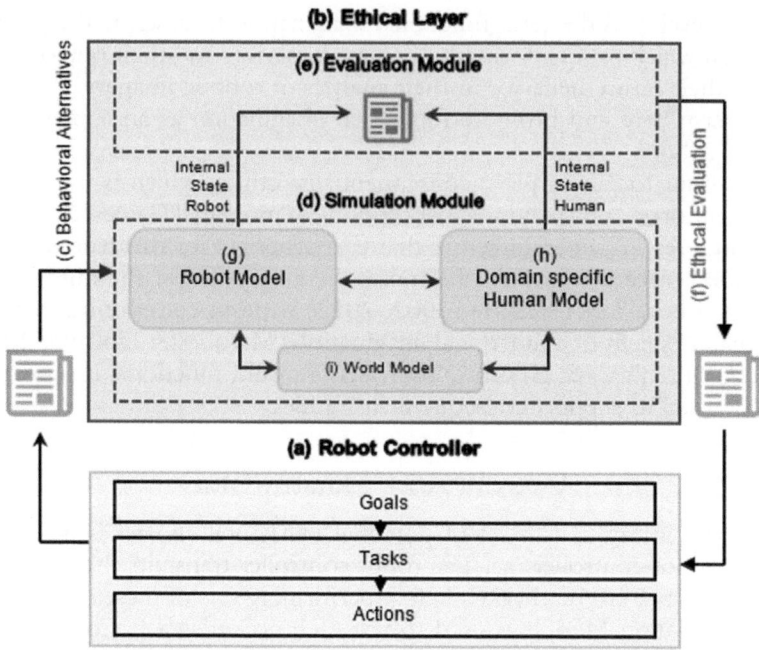

Fig. 3.1 Designing Ethical Robotic Interventions with Feedback (Using Simulation Theory of Cognition (Adapted from Vanderelst and Winfield (2018))

Proposition 1: ASD individuals and individuals with other learning disorders feel less intimidated and more comfortable by interacting with a social robot, and learn more efficiently than they would feel/learn with a human.

Proposition 2: Social robots can result in better educational, learning and interpersonal interactions / outcomes for ASD individuals and individuals with learning disabilities.

In the treatment and education of young people with autism, there are a number of potential uses for social robots. For instance, Frano Petric proposed a protocol to help with autism diagnosis. This protocol is based on four tasks extracted and modified by means of humane robots from the Autism Diagnostic Observation Schedule (ADOS). Previous researchers believe that by taking this approach, ASD diagnosis can be clearly defined, improving reliability and precision. Existing treatment research has been divided into three categories: the use of robots to (a) increase student involvement and motivation; (b) elicit behaviors; and (c) model, teach, and/or practice skills (Silvera-Tawil et al., 2018). These outcomes differ depending on the intervention method used, the robot used, and the magnitude of the child's symptoms. Hence,

Proposition 3: *When ASD individuals and individuals with other learning disorders / disabilities engage with ethical-social robots in educational / academic settings, the resulting human-robot interaction (HRI) leads to better student motivation and engagement; better social and interpersonal behaviors; and better overall learning and academic performance.*

DEVELOPMENT OF ETHICAL CURRICULUM COURSE DESIGN USING SOCIAL ROBOTICS

To ensure a diverse population that includes autistic people that can fully participate in a future society, we must prepare all students to build and critique emerging technologies such as artificial intelligence (AI). Classrooms are critical spaces for teaching students these skills, but there are few AI curricula developed for and used by teachers who teach Kindergarten through 12th grade.

In a previous study for middle school teachers that wanted to incorporate an ethical artificial intelligence (AI) lesson plan for their students, they created the 'How to Train Your Robot: AI and Ethics Curriculum' (Williams & Breazeal, 2020). In this study, we will use their curriculum as a guide to incorporate a new and improved ethical curriculum for high school and university students.

Many AI interventions have programming activities that allow students to experiment with AI and gain knowledge about it. 'The How to Train Your Robot: A Middle School AI and Ethics Curriculum' made use of tools from the Machine Learning for Kids and Teachable Machine projects to allow students to learn by doing. Non-Programming activities were also used in interventions to help students engage with AI in a variety of contexts. To help students gain new insights on AI before programming it, they used Payne's metaphors, like comparing algorithms to recipes, and non-programming activities, such as building a biased cat-dog classifier.

Although robotics education was not their primary focus, they included a physical robot in the course to encourage student learning and participation through embodied interaction. Physical objects can make concepts more realistic and thus easier to grasp, according to related fields such as mathematics and computer science education. More specifically, robots can help develop an appropriate hands-on experience the same results were observed by previous researchers that found benefits from incorporating robotics into their AI classrooms. In an undergraduate AI course, Kumar et al. (2019) used a LEGO robotics course to physically locate AI algorithms, providing another medium for understanding them beyond code and equations. The disadvantage of teaching AI with physical robots was that robots take a long time to physically maintain and do not always behave as one would expect (Kumar 2004; Williams & Breazeal, 2020). This is where that ethical layer inside the social robot will come into play alongside an ethically cultivated curriculum.

It is essential to classify robots due to the large variety of different types of social robots. In Appendix 1, you will find the characteristics of social robots, which are highlighted by differentiating between technical and social dimensions. The physical presence and therefore the design of the robot, from abstract to human, is the first technical dimension to distinguish a social robot (Guggemos et al., 2020). The robot's degree of autonomy, which ranges from remote control to complete autonomy, is the second technical dimension.

The social dimensions (Breazeal, 2003; Guggemos et al., 2020) depict the development stage of the robot interaction model on the one side and social embedding in the environment on the other. The interaction can be evocative and passive, or it can be sociable and proactive. The social embedding dimension places a greater emphasis on social conduct and incorporation into the environment. It can range from simple perception and response to the social environment to a fully socially intelligent robot (Fong et al., 2003; Guggemos et al., 2020).

We will be using a heuristic development approach in our framework to create solution-oriented applications for practical use with limited resources. In Appendix 2, you will find the conceptual framework for the development of use cases with social robots. The context imposes numerous constraints on the development team when creating a use case. As a first step, the development team should consider the context and document it. Later, the context implicitly defines the scope of action. The project must meet different requirements depending on the educational goal and setting. The development team must meet these requirements while also deliberately evaluating their know-how and anticipating the know-how and attitude of the future consumers.

The current infrastructure, available technology, and technological frontier should all be taken into account. In this context, integrating and utilizing existing services (such as text-to-speech services) may be preferable to in-house creation, as in-house development can be extremely costly. Finally, each development team must stay within budget and adhere to legal and ethical constraints (e.g., data protection policies), which impact what use cases may be implemented and which may not.

The context as a whole limit the types of applications that can be used. At the same time, the context establishes the parameters within which people and machines can act.

DISCUSSIONS AND CONCLUSIONS

The cognition simulation theory is the mind theory which enables the most rich depictions. It claims that thinking takes place as an engagement with the outer world through the same cognitive processes. Experiences are thus represented by particular cases and recombined by the perception, action, and emotional systems of the brain.

Actions are hidden when thinking and thought to provide sensory inputs that stimulate additional actions through associative brain mechanisms. Thinking requires building a physical representation of the environment rather than an abstract symbol. Instead, experience is thought to be represented by concrete events and recombined with the perception, action, and emotional system of the brain.

In the proposed ethical-social robotics course for ASD individuals we will incorporate a physical robot to promote learning and participation of students through personalized interaction. Physical things can make concepts more realistic and hence simpler to absorb in the domains of mathematics and information technology education. In particular, robots can help to produce a suitable practical experience with the same results that past scientists have seen that the robot integration has benefitted in their AI courses.

Social robotics course can improve ASD individuals' learning and academic performance by giving them the freedom to express themselves without feeling judged. This will encourage ASD individuals to continue working and do better.

APPENDIX 1

See Table 3.1.

Table 3.1 Classifying the characteristics of social robots

	DIMENSIONS	CHARACTERISTICS			
Technical Dim.	Design of physical presence	Iconic, abstract design of robots (e.g. Jibo)	Animal-like robots (e.g. Robear as care robot)	Humanoid robots (e.g. Pepper, Nao)	Android, human-like design (e.g. Sophia)
	Control and autonomy	Remotely controlled, e.g. telepresence representative	Semi-autonomous: local AI and webservices	Autonomous systems with local AI that allows them to interact independently	
Social Dim.	Interaction model	Socially evocative (e.g. rely on human engagement)	Social interface (e.g. social behavior is modeled)	Socially receptive (e.g. learning skills by imitation)	Sociable (e.g. proactive engagement)
	Social embedding	Socially situated (e.g. perceive and react to social environment)	Socially embedded (e.g. interact with social environment)	Socially intelligent (human cognition and social competence)	

Note Draws on the work of Breazeal (2003), Duffy (2003), Fong et al. (2003), Belpaeme et al. (2018), and Baraka et al. (2020) as cited in Guggemos et al. 2020

APPENDIX 2

See Table 3.2.

Table 3.2 Conceptual framework for the development of use cases with social robots

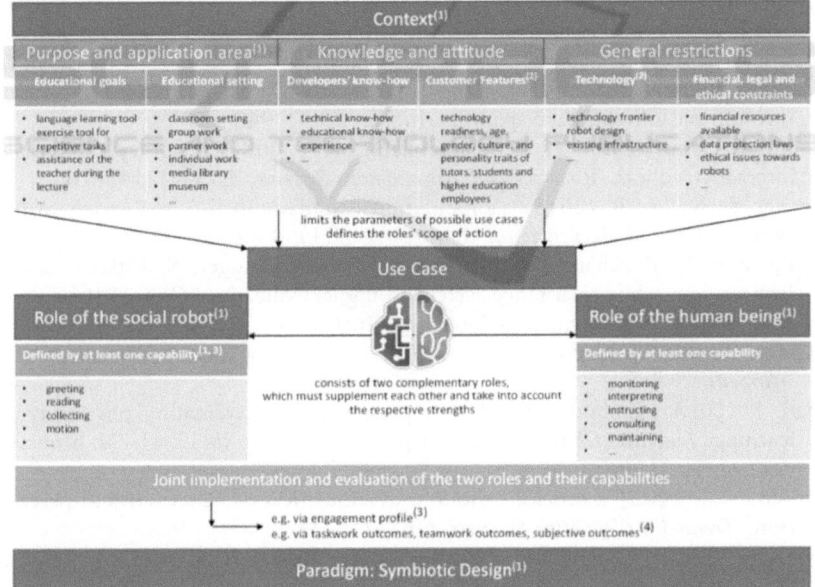

Note (1) Baraka et al. (2020), (2) Belanche et al. (2020), (3) Cooney and Leister (2019), (4) You and Robert (2018) as cited in Guggemos et al. 2020

References

Anderson, M., & Anderson, S. L. (2007). Machine ethics: Creating an ethical intelligent agent. *AI Magazine, 28*(4), 15–15.

Baraka, K., Alves-Oliveira, P., & Ribeiro, T. (2020). An extended framework for characterizing social robots. In *Human-robot interaction* (pp. 21–64).

Barr, D., Harrison, J., & Conery, L. (2011). Computational thinking: A digital age skill for everyone. *Learning & Leading with Technology, 38*(6), 20–23.

Beeson, P. M., & Robey, R. R. (2012). Academy of Neurologic Communication Disorders and Sciences (ANCDS) Aphasia Treatment Evidence Tables.

Bers, M. (2010). The TangibleK robotics program: Applied computational thinking for young children. *Early Childhood Research and Practice, 12*(2), 1–20.

Boccanfuso, L., & O'Kane, J. M. (2011). CHARLIE: An adaptive robot design with hand and face tracking for use in autism therapy. *International Journal of Social Robotics, 3*(4), 337–347.

Breazeal, C. (2003). Toward sociable robots. *Robotics and Autonomous Systems*, 42(3–4), 167–175.

Diagnostic and Statistical Manual of Mental Disorders. (2000). *Fourth Edition (DSM-IV) (2000)*. American Psychiatric Publishing Inc.

Fletcher-Watson, S., Adams, J., Brook, K., Charman, T., Crane, L., Cusack, J., & Pellicano, E. (2019). Making the future together: Shaping autism research through meaningful participation. *Autism*, 23(4), 943–953.

Fong, T., Nourbakhsh, I., & Dautenhahn, K. (2003). A survey of socially interactive robots. *Robotics and Autonomous Systems*, 42(3–4), 143–166.

Grover, S., & Pea, R. (2013). Computational thinking in K–12: A review of the state of the field. *Educational Researcher*, 42(1), 38–43.

Guggemos, J., Burkhard, M., Seufert, S., & Sonderegger, S. (2020). Social Robots as teaching assistance system in higher education: Conceptual framework for the development of Use cases. In *CSEDU (1)* (pp. 125–132).

Haines, W. (2015). Consequentialism. Internet encyclopedia of philosophy. *Iep. utm. edu.*

Han, I. (2013). Embodiment: A new perspective for evaluating physicality in learning. *Journal of Educational Computing Research*, 49(1), 41–59. https://doi.org/10.2190/EC.49.1.b

Hesslow, G. (2002). Conscious thought as simulation of behaviour and perception. *Trends in Cognitive Sciences*, 6(6), 242–247.

Hesslow, G. (2012). The current status of the simulation theory of cognition. *Brain Research, 1428*, 71–79.

Kennedy, J., Baxter, P., & Belpaeme, T. (2014). Children comply with a robot's indirect requests. In *Proceedings of the 2014 ACM/IEEE international conference on Human-robot interaction*, March 2014, 198–199.

Kopcha, T. J., McGregor, J., Shin, S., Qian, Y., Choi, J., Hill, R., & Choi, I. (2017). Developing an integrative STEM curriculum for robotics education through educational design research. *Journal of Formative Design in Learning, 1*(1), 31–44.

Kumar, A. N. (2004). Three years of using robots in an artificial intelligence course: lessons learned. *ACM Journal on Educational Resources in Computing*, (JERIC), 4(3), 1–15.

Kumar, R., Mandalika, A., Choudhury, S., & Srinivasa, S. (2019). Lego: Leveraging experience in roadmap generation for sampling-based planning. In *2019 IEEE/RSJ International Conference on Intelligent Robots and Systems (IROS)*, IEEE, November 2019, 1488–1495.

Mackworth, A. K. (2011). *Architectures and ethics for robots: Constraint satisfaction as a unitary design framework*. Cambridge University Press.

Marques, H. G., & Holland, O. (2009). Architectures for functional imagination. *Neurocomputing, 72*(4–6), 743–759.

Moor, J. H. (2006). The nature, importance, and difficulty of machine ethics. *IEEE Intelligent Systems, 21*(4), 18–21.

Royakkers, L., & van Est, R. (2015). A literature review on new robotics: Automation from love to war. *International Journal of Social Robotics, 7*(5), 549–570.

Sengupta, P., Kinnebrew, J., Basu, S., Biswas, G., & Clark, D. (2013). Integrating computational thinking with K-12 science education using agent-based computation: a theoretical framework. Education and Information Technologies, 18, 351–380. 10. 1007/s10639–012–9240-x.

Silvera-Tawil, D., Bradford, D., & Roberts-Yates, C. (2018, August). Talk to me: The role of human-robot interaction in improving verbal communication skills in students with autism or intellectual disability. In *2018 27th IEEE International Symposium on Robot and Human Interactive Communication (RO-MAN)* (pp. 1–6). IEEE.

Vanderelst, D., & Winfield, A. (2018). An architecture for ethical robots inspired by the simulation theory of cognition. *Cognitive Systems Research, 48*, 56–66.

Williams, R., & Breazeal, C. (2020). *How to train your robot: A middle school AI and ethics curriculum.* IJCAI. https://robots.media.mit.edu/wp-content/upl oads/sites/7/2020/07/ijcai_2020.pdf

Wilson, M. (2002). Six views of embodied cognition. *Psychonomic Bulletin & Review, 9*(4), 625–636.

Wing, J. M. (2006). Computational thinking. *Communications of the ACM, 49*(3), 33–35.

CHAPTER 4

Recalibrating Anthropomorphized Robotic Interactions During COVID-19: Understanding Human Robotic Interactions

Ndifreke Akpan and Anshu Saxena Arora

Abstract Human behavior and emotional intelligence add to the backdrop for social interactions. Robots are becoming more human-like with every new update, remodel, and release. In a time where human interaction is minimized, anthropomorphic recalibrations become a baseline for new designs. The COVID-19 pandemic has forced technology engineers to re-focus. Lately, where single adults are less likely to engage in new social connections, human–robot interactions have become the norm. All cultures and demographics alike are affected. Robot/technology with human personality traits, facial recognition, inflection in tone, and reading temperatures are all part of a demand in current upgrades. This research

N. Akpan · A. S. Arora (✉)
University of the District of Columbia, Washington, DC, USA
e-mail: anshu.arora@udc.edu

N. Akpan
e-mail: ndifreke.akpan@udc.edu

© The Author(s), under exclusive license to Springer Nature Switzerland AG 2022
A. S. Arora et al. (eds.), *Managing Social Robotics and Socio-cultural Business Norms*, International Marketing and Management Research,
https://doi.org/10.1007/978-3-031-04867-8_4

43

dives into the degrees of varying changes and adjustment timeframes when robots are introduced to new experiences. The research focuses on human personality dimensions in human–robot interaction situations and examines the relationships between human personality traits and robotic anthropomorphism across different demographics and emotional changes regarding overall robot likeability and acceptance.

Keywords Anthropomorphic recalibrations · COVID-19 pandemic · Human–robot interaction · Social robotics · Cultures · Demographics · Human personality traits · Robot likeability · Robot acceptance

INTRODUCTION

The goal of this paper is to provide an extensive overview of artificial intelligence (AI) and human–robot interaction (HRI), anthropomorphism, and social robotics amid the coronavirus disease 2019 (COVID-19) pandemic. The research illustrates how anthropomorphized robots aid in medical care and comfort, reducing the spread of illness, and reducing social anxiety for those left without human contact. The research addresses the following questions:

1. Does AI address the issue of crisis management?
2. How can anthropomorphized robots aid in comforting those who have no human aid?
3. Can robots substitute for humans to reduce anxiety from loneliness?

The research makes several contributions. First, there is an abundance of scientific works related to AI and social robotics, however, because of the newness of the pandemic, there is very little conclusive research on robotic anthropomorphism and COVID-19. A virus of this magnitude that can be transmitted through literally breathing in the same space as someone has created a need for social distancing mandates. The use of technology has become a pivotal resource in health care. Initially, referenced as future aid instruments, they've now come to the forefront and are categorized as essential since mandates are for a significant reduction in human contact (Clipper, 2020). Second, human personalities, expressions, and inflections are all part of the natural communication language. In our expressions, we tend to be a little more extroverted, meaning showing readable expressions that are easily noticed through sight. These behavioral differences create an ideology of the world around us. It affects how we view peers, which now include robots. Research has showed that robots are more apt to anthropomorphize at the strength of extroversion. One conclusion was that participants swayed the robots they interacted with through their own personality dimensions (Kaplan, 2019).

This paper is organized in the following manner: Sect. 1 provides an overview of the use of AI and social robotics for crisis management. Section 2 provides literature review on social robotics, anthropomorphism and human personality traits. Section 3 presents a conceptual framework of use of HRI to mitigate loneliness during global crisis management. Section 4 discusses the managerial implications of social robotics for future.

Theoretical Background

Does AI Address the Issue of Crisis Management?

The social planet has been affected by COVID-19, resulting in a global pandemic. Stay at home orders, work-from-home, virtual learning for schools, and video only doctor's appointments are a few pivoting measures used to combat the crisis. Artificial Intelligence, in lieu of human interaction has emerged as a forefront solution in various processes. Vint Cerf, dubbed as the Father of the Internet, VP and CIE at Google, also a survivor of COVID-19 expresses that AI technology will have a "prominent role" in restructuring our civilization post COVID-19. He goes on to declare it an "awesome responsibility!" (Cerf, 2020, p. 7). According to the CDC, contact tracing is the continual identification, monitoring, and support of a confirmed or probable case's close contacts who have been exposed to, and possibly infected with, the virus. The infected patient's identity is not discussed with contacts, even if asked.

There have been several suggestions for the best method to obtain this information. One such solution has been the use of smart phones in proximity tracking. According to the CDC, there are advantages as well as disadvantages to this solution. One such advantage, as can be expected via technology, is its ability to provide more comprehensive mobility history;

the infected person would most accurately be able to recall their movements and convey accurate information to medical officials. A popularized disadvantage, however, is data security and unauthorized access through hacking and compromised confidentiality.

The use of AI increases urgency to save lives during an epidemic but so does the urgency of ethics. The rapid rate in which solutions are demanded creates its own set of risks. Concerns loom amidst the challenge of providing solutions by gathering an enormous amount of personal data. Societal expectations in an age where AI resolves problems seamlessly have created a need for immediate solutions. Everything is delivered at the snap of a finger. Information is accessed across the globe in a matter of seconds. The data collected and processed during the past 8 months doesn't yield enough time to access solutions as in times past. Research for a virus of this magnitude would typically take years, even decades to be conducted. We are now in a space where resolution has been demanded in less than a year. The efficacy of data collection is then questioned as rapid results are demanded and a vetting process has not been established for newly collected data. In AI simulations that result in no discrepancies the predictability of its positive impact across diverse populations could still be hard to determine. In addition to the accuracy of data pulled, the government and public officials' transparency has added to the diminishing trust of the public. The way this plays out will depend on what the public views as a responsible use of AI in addressing COVID-19 (Tzachor et al., 2020).

How Can Anthropomorphized Robots Aid in Comforting Those Who Have No Human Aid?

In the wake of the COVID-19 pandemic social isolation in elderly community is increasingly become a common occurrence. Families, for fear of transmitting COVID-19 opt to distance themselves from their aging loved ones. Younger/healthier adults can carry the virus asymptomatically, then passing it along to the more at-risk population. A safety measure has been to eliminate contact completely. This leaves the elderly in a disproportionately isolated condition. More than ever, they have resorted to almost solely relying on communication thorough virtual/digital means.

Anthropomorphized brands and objects have been met with more exceptional significance in recent years, according to research. Evidence demonstrates that products keenly involved in their own consumption are handled differently than those regarded as just objects. (Yang, Aggarwal, et al., 2020; Yang, Nelson, et al., 2020). The desire to anthropomorphize creates a sense of ownership and belonging that is not otherwise present in inanimate objects. It is worth noting here that the author has had personal experiences with naming vehicles and referring to mechanical problems as 'misbehaving'. Understanding the reasons behind comfort and assurance in robots increases as they become more prevalent in our society. HRI can be attributed to the individual robots or the personalities of the humans they service (Kaplan et al., 2019). People who are considered more technologically savvy have less resistance to navigating robotic technology. In such cases, anthropomorphizing comes easier. Using robots for communicating, assisting, and scheduling becomes more second nature. Social psychology proves that discernment is enhanced after repeated interactions (Paetzel et al., 2020).

In the case of a pandemic such as COVID-19, previous studies on social robotics to aid in comforting the loneliness of loved ones, such as elderly parents, were undertaken. The interest was in finding how robot anthropomorphism and social robotics could alleviate and avoid all together feelings of anaclitic depression. Standard communication using phones, non-expressive mediator robots, and expressive mediator robots were compared. The study was conducted on 741 elderly people. Social mediator robots were used to create ease of self-disclosure in seniors who would otherwise maintain discretion of their true feelings. The study showed that seniors, for varying reasons, kept certain information to themselves. A second, parallel study presented more detailed analysis of the type of robots paired with a respective participant. The results concluded that the robots regarded as "more expressive" were those with social expressions, more anthropomorphized. This caused the elderly participants to disclose more. The second study concluded that through HRI, matching robot personalities produces more confidence on the human's performance (Noguchi et al., 2020) see Fig. 4.1.

Can Robots Substitute for Humans to Reduce Anxiety from Loneliness?

COVID-19 creates feelings of loneliness among those in isolation. The socially deadly virus has led researchers to analyze its effect on human social isolation. Social companion robots have been engaged to reduce and in cases where possible, eliminate feelings of loneliness.

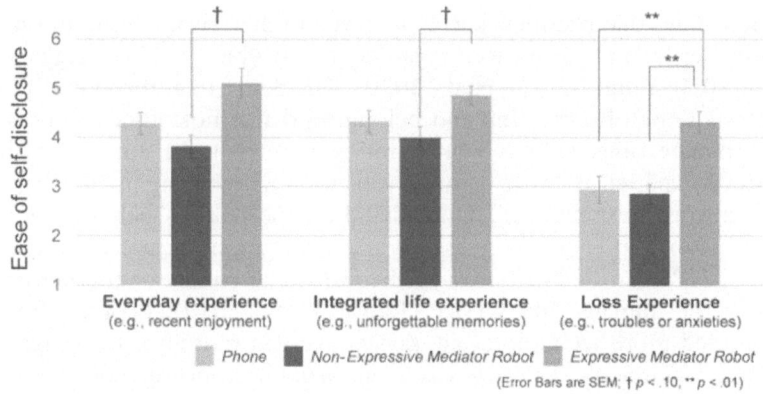

Fig. 4.1   Ease of self-disclosure through communication media (Noguchi et al., 2020)

A study conducted with 595 online users provided insight to the use and necessity of a small companion robot by the name of Vector (see Fig. 4.2). The contributions of this study are bipartite. The first proposition is that robots can mitigate lonely feelings; and the second is that the types of helpful interactions through which the robots assist can cause uplifting changes and improvements in the well-being of the participants. The study used netnography, a research method originating in the description of the customs of individuals that examine online cultures and life. Data are gathered through online comments, blog posts and unsolicited customer reviews (Kozinets, 2002, p. 62). This data was collected between January and June of 2020, the height of the pandemic. Vector, a small personal companion robot produced by Anki, had many users. More users purchased and reviewed the robot during the beginning of the pandemic. Described by Anki as "your buddy" (https://www.digitaldreamlabs.com/collections/vector-products), Vector was researched via signal words and semantic assessments (Heinonen & Medberg, 2018).

Vector was used in various ways, aiding with homework for students, as a personal assistant for adults, meteorology, cooking timers, and searching random knowledge through its interconnection with Amazon's Alexa. Vector has also been used for social connectivity, sharing recipes, and entertainment to resolve boredom. Users also display feelings of social identity to Vector, the researchers reported word uses such as "love,

Fig. 4.2 Photo of vector (Retrieved from https://www.walmart.com/)

adore" or "fantastic". Vector has been affectionately referred to as their "child" or family member (Odekerken-Schröder, et al., 2020). These findings express that robot can indeed mitigate human loneliness, but the long-term effects are inconclusive. This would be an on-going study as we're still less than a year into the study/use of robots to substitute for human interaction.

CONCEPTUAL FRAMEWORK

Hedonic well-being is equated with pleasure and happiness and often operationalized as satisfaction and positive affect or the absence thereof (Diener, 2012; Diener & Lucas, 1999). The eudaimonic form defines well-being along a set of dimensions that promote meaning and self-realization (e.g., environmental mastery, personal growth and positive social relations; Ryff, 1989) to advocate fully functioning individuals (Rogers, 1963) (see Fig. 4.3).

As robots vary in type and functionality, users may find that not all robots or AI are created equally. Varying types of HRI provide different

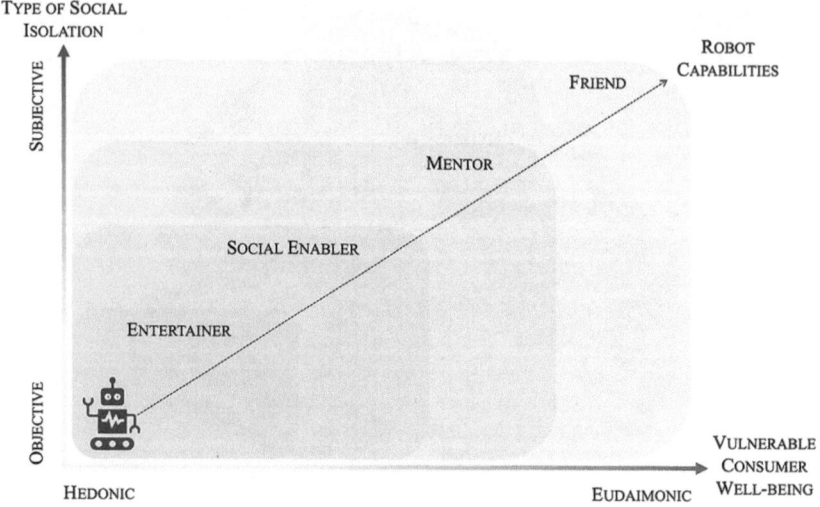

Fig. 4.3 Robot capabilities during COVID-19 (2020 Adapted from Henkel et al.,)

response outcomes. Through our conceptual framework (see Fig. 4.4), we see that the robots with immediate responses, typically something preprogrammed to behave to set commands, only provide temporary relief. Thus, the idea of being entertained is accompanied by a satisfactory timeline, after which humans are too distracted to need robots. For satisfying both (human) utilitarian and hedonic needs and overall well-being of a person, the robot would require more anthropomorphizing and HRI. The robot would have to anticipate need, offer resolutions, and maintain data that makes it invaluable.

Mitigating loneliness during a pandemic qualifies a range of robots. The type of robot and its use is most valuable to the need of the human. Earlier in the paper, we examined the use of moderator robots with the elderly. The human's desire to disclose more information to more anthropomorphized robots conveyed a comfort in communicating with objects that most feel human-like. Continuous HRI eases the burden of isolation but also increases feelings of comfort in most humans (Henkel et al., 2020).

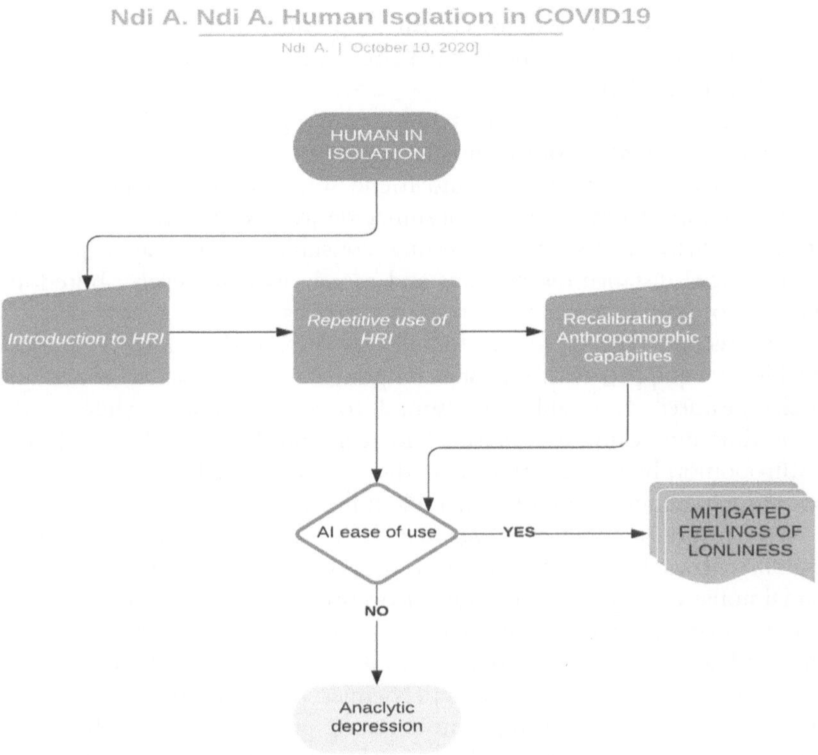

Fig. 4.4 Our conceptualization: Flow chart diagraming benefits of AI in HRI to mitigate loneliness

Managerial Implications of Social Robotics

Good health is associated with a healthy social life. People dealing with terminal ailments, and emotional disabilities experience an adverse effect on their social life and overall health. This not only affects the value of their life experiences but also their mortality. Socially Assistive Robots (SARs) are designed to aid users in areas of social as opposed to physical connections. The healthcare usage of SARs functions to conduct interview screenings, document symptoms, sort medications, aid in transitions through therapies, remove stress, and aid in social communications among humans (Chita-Tegmark & Scheutz, 2020). Research finds that vulnerable consumers require cheering up and robot interpretations of their emotional state can be learned over time and continuous exposure (Henkel et al., 2020).

Additionally, the study concludes, among four types of robots researched, one has more futuristic impact than the others. The entertainer, social enabler, mentor, and friend were the robot types studied. The robot most described as consumer desired is the *entertainer*. This robot is characterized by addressing consumers who may experience minor social distance discomforts and merely used to resolve boredom. This robot only executes monotonous exercises.

The second type of robot studied was the *social enabler*. The social enabler is equipped with sensory components which react to touch and social gestures. It is able to respond to social contacts while at the same time mimicking expressions (Adalgeirsson & Breazeal, 2010). The anthropomorphized capabilities of this robot type allow it to mediate social interactions for more vulnerable humans.

The future-oriented type *mentor* serves as a transformative robot. Initially created for overcoming disproportions of hedonic but more eudaimonic feelings. Social isolation occurrences can include healthcare professionals. In which case robots can autonomously engage through human-like competences. Mentor robots will then have the capacity to exemplify teachers and instructors (Niemiec & Ryan, 2009). Prolonged use with mentor robots can promote lasting positive effects. Robots with more human-like gestures head, eye, and mouth animations are preferred to those that are immobile (Zinina et al., 2020).

Through quasi-social exchanges, robots that are referred to as *friend*, can lessen unfavorable outcomes by building relationships through a sense of touch and proprioception; This would reduce negative impacts of social isolation (Henkel et al., 2020).

FUTURE RESEARCH

Notwithstanding, the creation of a robot with AI calibrated for empathy would be incredibly impactful. Social robotics in the future benefit from greater research into the use of robotics to respond to infectious diseases. Promoting collaboration between technology developers and infectious disease scientists, and adequate sponsorships being ahead of the next pandemic would be a thing of the past (Yang, Aggarwal, et al., 2020; Yang, Nelson, et al., 2020). The use of social robots such as NAO, Furhat, and Missy Robotics, along with robotic voice-assistants like Alexa, Siri, and Vector can help society into combatting the effects of social isolation. Anthropomorphizing resulting from curating empathetic responses through HRI can bring us into a world where the option to eradicate loneliness exists.

REFERENCES

Adalgeirsson, S. O., & Breazeal, C. (2010), *"MeBot: a robotic platform for socially embodied telepresence"*, 2010 5th ACM/IEEE International Conference on Human-Robot Interaction (HRI), presented at the 2010 5th ACM/IEEE International Conference on Human-Robot Interaction (HRI), IEEE, pp. 15–22, https://doi.org/10.1109/HRI.2010.5453272.

Cerf, V. G. (2020). Implications of the COVID-19 pandemic. *Communications of the ACM, 63*(6), 7–7.

Chita-Tegmark, M., & Scheutz, M. (2020). Assistive robots for the social management of health: A framework for robot design and human–robot interaction research. *International Journal of Social Robotics*, 1–21.

Clipper, B. (2020). The influence of the COVID-19 pandemic on technology: Adoption in health care. Nurse Leader.

Diener, E., & Lucas, R. E. (1999). Personality and subjective wellbeing. In D. Kahneman, E. Diener, & N. Schwartz (Eds.), *Wellbeing: The Foundations of Hedonic Psychology* (pp. 213–229). Russell Sage Foundation.

Diener, E. (2012). New findings and future directions for subjective well-being research. *American Psychologist, 67*(8), 590–597. https://doi.org/10.1037/a0029541

Heinonen, K., & Medberg, G. (2018). Netnography as a tool for understanding customers: Implications for service research and practice. *Journal of Services Marketing, 32*(6), 657–679. https://doi.org/10.1108/JSM-08-2017-0294

Henkel, A. P., Čaić, M., Blaurock, M., & Okan, M. (2020). Robotic transformative service research: Deploying social robots for consumer well-being during Covid-19 and beyond. *Journal of Service Management*.

Kaplan, A. D., Sanders, T., & Hancock, P. A. (2019). The relationship between extroversion and the tendency to anthropomorphize robots: A Bayesian analysis. *Frontiers in Robotics and AI, 5*, 135.

Kozinets, R. V. (2002). The field behind the screen: Using netnography for marketing research in online communities. *Journal of Marketing Research, 39*(1), 61–72. https://doi.org/10.1509/jmkr.39.1.61.18935

Niemiec, C. P., & Ryan, R. M. (2009). Autonomy, competence, and relatedness in the classroom: Applying self-determination theory to educational practice. *Theory and Research in Education*, 7(2), 133–144. https://doi.org/10.1177/1477878509104318

Noguchi, Y., Kamide, H., & Tanaka, F. (2020). Personality traits for a social mediator robot encouraging elderly self-disclosure on loss experiences. *ACM Transactions on Human-Robot Interaction (THRI)*, 9(3), 1–24.

Odekerken-Schröder, G., Mele, C., Russo-Spena, T., Mahr, D., & Ruggiero, A. (2020). Mitigating loneliness with companion robots in the COVID-19 pandemic and beyond: an integrative framework and research agenda. *Journal of Service Management*.

Paetzel, M., Perugia, G., & Castellano, G. (2020, March). The persistence of first impressions: The effect of repeated interactions on the perception of a social robot.

Rogers, C. R. (1963). The actualizing tendency in relation to 'motives' and to consciousness. In N. E. Lincoln, M. R. Jones (Ed.), *Nebraska Symposium on Motivation*, University of Nebraska Press, Vol. 11, pp. 1–24.

Ryff, C. D. (1989). Happiness is everything, or is it? explorations on the meaning of psychological well-being. *Journal of Personality and Social Psychology*, 57(6), 1069–1081. https://doi.org/10.1037/0022-3514.57.6.1069

Tzachor, A., Whittlestone, J., & Sundaram, L. (2020). Artificial intelligence in a crisis needs ethics with urgency. *Nature Machine Intelligence*, 2(7), 365–366.

Yang, G. Z., Nelson, B. J., Murphy, R. R., Choset, H., Christensen, H., Collins, S. H., ... & Kragic, D. (2020a). Combating COVID-19—The role of robotics in managing public health and infectious diseases.

Yang, L. W., Aggarwal, P., & McGill, A. L. (2020). The 3 C's of anthropomorphism: Connection, comprehension, and competition. *Consumer Psychology Review*, 3(1), 3–19.

Zinina, A., Zaidelman, L., Arinkin, N., & Kotov, A. (2020). Non-verbal behavior of the robot companion: A contribution to the likeability. *Procedia Computer Science*, 169, 800–806.

CHAPTER 5

Robotic Anthropomorphism and Intentionality Through Human–Robot Interaction (HRI): Autism and the Human Experience

Andrew Sammonds, Anshu Saxena Arora, and Amit Arora

Abstract This research focuses on robotic anthropomorphism and how it impacts the learning environment of students with autism spectrum disorder (ASD). ASD students show a greater interest in anthropomorphic characteristics in robots. Social interaction between robots and students by employing anthropomorphism degrees in a robot's physical design and behavior has boosted productivity in ASD students. As robots

A. Sammonds · A. S. Arora (✉) · A. Arora
University of the District of Columbia, Washington, DC, USA
e-mail: anshu.arora@udc.edu

A. Sammonds
e-mail: andrew.sammonds@udc.edu

A. Arora
e-mail: amit.arora@udc.edu

© The Author(s), under exclusive license to Springer Nature Switzerland AG 2022
A. S. Arora et al. (eds.), *Managing Social Robotics and Socio-cultural Business Norms*, International Marketing and Management Research, https://doi.org/10.1007/978-3-031-04867-8_5

enter our social space, we will inherently impose our interpretation on their actions, similar to the techniques we employ in rationalizing, for example, a pet's behavior. This propensity to anthropomorphize is not seen as a hindrance to social robot development but rather a helpful mechanism that requires careful examination and employment in social robotics research. Specifically, this chapter examines social-cognitive intelligence in relation to artificial intelligence, emphasizing privacy protections and ethical implications of HRI, while designing robots that are ethical, cognitively, and artificially intelligent, as well as human-like in their social interactions.

Keywords Robotic Anthropomorphism · Robotic Intentionality · Social Cognition · Autism Spectrum Disorder · Human–Robot Interaction · Humanoid Robots · Social Robotics · Human–Robot Interaction (HRI)

INTRODUCTION

Almost all the famous childhood stories use anthropomorphism in some way. Those stories, in most cases, feature human characters interacting with non-human characters. Social robots have a special relationship with anthropomorphism, which they consider neither a cognitive error nor a sign of immaturity (Damiano & Dumouchel, 2018). Instead, it considers that this common human tendency, which is supposed to have evolved because it favored cooperation among early humans, can be used today to facilitate social interactions between humans and a new type of collaborative and interactive agents—social robots (Damiano & Dumouchel, 2018). This approach leads social robots to focus research on engineering robots that activate users' stereoscopic projections. The goal is to give robots a "social presence" and "social behaviors" that are credible enough for human users to engage in comfortable, long-term relationships with these machines (Damiano & Dumouchel, 2018). This choice of "applied anthropomorphism" as a research method exposes the artifacts produced by social robots to moral condemnation: social robots are judged as a "cheating" technology because they generate in users the illusion of mutual social and emotional relationships (Damiano & Dumouchel, 2018). This chapter takes a position in this debate, developing a series of arguments relevant to the philosophy of mind, cognitive science, and robotic artificial intelligence and asking what social robots can teach about anthropomorphism.

The Social Cognitive Theory (SCT) examines the influences of individual experiences, the actions of others, and environmental factors on health behaviors. Through instilling expectations, promoting self-efficacy, using observational learning, and using other reinforcements to change behavior, Social Cognitive Theory offers opportunities for social support. Exploration shows that individuals with ASD are especially powerless against forlornness, and hence the humanizing of non-human specialists may work as a social outlet of sorts. For example, grown-ups with a severe level of ASD-related qualities were discovered to be the same as controls in their craving for friendship yet detailed altogether higher evaluations of forlornness which they credited to their absence of social agreement (Jobe & Williams White, 2007). Proof of less informal organizations (Mazurek, 2014), alongside an expanded impression of the self as a helpless social entertainer (Vickerstaff et al., 2007), may add to the raised degrees of social nervousness present inside the populace (for an audit, see MacNeil et al., 2009). As social contrasts may seclude those with ASD from peers and result in adverse results, humanizing non-human substances may consider social commitment with less passionate danger. Along these lines, collaborations with human characters may turn out to be more socially inspiring.

There is a dearth of research in the field of robotic anthropomorphism and intentionality in social robotics and the way these concepts can impact social-cognitive behavior of ASD individuals. Human–robot interaction (HRI) studies have shown exciting yet preliminary benefits for individuals with autism spectrum disorder, including increased engagement in tasks, increased levels of attention, and novel social behavior, such as joint attention. Despite the excitement generated by these studies within the robotics community and media attention, the results have received relatively little attention from the clinical community; clinicians tend to view HRI for autism as a trial or an experiment. Presently, research in advanced social mechanics and HRI is investigating the impact of ascribing deliberateness to robots and the conduct boundaries of the robot that most proficiently instigate this. In investigating the impact that crediting deliberateness has on friendly communication, members of a test are occasionally persuaded that they are collaborating either with a pre-customized machine (e.g., Wykowska et al., 2014; Özdem et al., 2017) or with another human (who normally has wants and convictions).

In some different examinations, members are first presented to various specialist types (e.g., human, humanoid robot, non-humanoid robot) and consequently are persuaded that they are connecting with one of them (e.g., Krach et al., 2008). Prompting a specific (e.g., deliberate) position through guidance control remains rather than strategies utilized in research that means to characterize the boundaries under which members unexpectedly expect the purposeful position. Here, the deliberate position is actuated through, for instance, the robot's look, discourse, or general conduct (e.g., Wykowska et al., 2015).

1. Can robots be perceived as "intentional" agents?
2. How can social robots facilitate student learning for individuals with autism spectrum disorder and other learning disorders/disabilities through robotic anthropomorphism and intentionality?
3. How can we change human behavior through or with human–robot interaction (HRI) situations?

The research makes the following contributions. First, it aims to develop a different perspective on social robots, as it explains how social robots can be scientific tools for examining human social cognition, particularly its flexibility. Secondly, it focuses on the importance of social robotics through anthropomorphism and intentionality and how it may improve social cognition for ASD individuals, and thirdly, we will consider the issue of adopting an intentional stance toward robots, discuss its relationship to other, lower-level mechanisms of social cognition, and evaluate methods to assess adoption of intentional stance. The main purpose of this research is to focus on social robotics' concepts of anthropomorphism and intentionality and how they can result in improved social cognition for ASD individuals. This chapter consists of four sections. First, we focus on defining and describing robotic anthropomorphism and intentionality in social robotics. Second, we examine how social-cognition traits of ASD individuals can be improved through robotic anthropomorphism and intentionality. Thereafter, we investigate the effects of robotic anthropomorphism and intentionality on robot likeability, and the subsequent effect of these HRI interrelationships on the success of HRI implementation. Next, we propose our Anthropomorphism—Intentionality—Social-Cognition framework and propose managerial implications of our framework on consumers and businesses in the context of social robotics.

Theoretical Background

A study examines the direct effects of deliberately adopting social media, deliberately acting as an independent variable, and caused by the misrepresentation of beliefs. At the same time, the functioning of robots is similar to experimental conditions. In line with this, further research is looking at ways in which the intentional state may be created automatically by robots' behavior (Terada et al., 2008; Yamaji et al., 2010). Where deliberate persuasive approaches themselves are the subject of research, measures to measure the effectiveness of this process are needed, and although the objective nature as a concept is defined as length (Dennett, 1971, 1987), measuring its acceptance presents a challenge. In addition, much of the literature on intentional architecture comes from the field of engineering and has a different approach to previously discussed research based on experimental psychology. Although the similarities are evident in the intentions and the overall term, the method and research questions in these different fields are often different. A prominent type of paradigm in HRI research in targeted mental states includes natural experiments and open conclusions (e.g., Terada et al., 2007; Yang et al., 2015). Participants in these studies are usually not given strict instructions on how to perform a particular task with the robot in question, but it leaves the connection naturally natural. Despite the similarities between this type of setup and the nature of the set of robotic platforms, the authenticity of the test is compromised in this way, and questions about which aspect of the robotic behavior leads to the acceptance of the target state remain unanswered.

Apart from discussing the circumstances in which the state of determination and the role played by the position played in balancing social perceptions of society separately, these conditions do not exist in isolation. This is illustrated by Pfeiffer et al. (2011), which showed that the personality values associated with the on-screen avatar in the public view test differ from the combination of the viewing behavior and the subject's expectations regarding the avatar. According to the task order, personality limitations increased when the avatar's visual performance seemed to be in line with the strategy that followed. Therefore, it is important to know that in the end, it is a combination of moral boundaries and human expectations for the integrated robot to inform human values. At present, this is rarely considered in an HRI study.

ASD is a progressive disorder characterized by poor social and interpersonal communication, in line with known and recurring behavioral limitations and interests (Lord et al., 1994). For example, children with ASD avoid physical contact, do not target people, do not show connection, do not show excitement or interest, and may spend many hours listing toys or investigating items (Rutter et al., 2003). Since ASD cannot be cured, some people with the condition need more expensive and more powerful, lifelong care and treatment, which encourages the development of social robots to help them and their caregivers. The emergence of social robots dedicated to ASD can be traced back to seminal research by Emanuel and Weir (1976) (see also Howe, 1983), where an electron-controlled computer, a tortoise-like robot (LOGO) that rotates on the ground, was used as a correction tool ASD boy. It was not until the late 1990s that many laboratories adopted this research topic (see Werry & Dautenhahn, 1999; Diehl et al., 2012; Begum et al., 2016; Ismail et al., 2019; review).

Until now, around 30 robots have been tested as ASD correction tools [e.g., Robota (Billard et al., 2007), FACE (Pioggia et al., 2007), Aibo (Francois et al., 2009), Charlie (Boccanfuso and O'Kane, 2011), NAO (Shamsuddin et al., 2012; Arora & Arora, 2020), GIPY-1 (Giannopulu, 2013), Pleo (Kim et al., 2013), KASPAR (Wainer et al., 2014), Jibo (Guizzo, 2015), Maria (Valadao et al., 2016), Sphero (Golestan et al., 2017), MINA (Ghorbandaei Pour et al., 2018), Leo (She et al., 2018), SAM (Lebersfeld et al., 2019), SPRITE (Clabaugh et al., 2019), Actroid-F (Yoshikawa et al., 2019) etc.]

Anthropomorphism is defined as "seeing the human in non-human forms" (Aggarwal & McGill, 2007). In the field of social robotics, creators and designers design robots with human (or living being) like characteristics to incite robotic anthropomorphism (e.g., Nao, the bipedal humanoid robot developed by SoftBank Robotics is used popularly in education and research). Uncanny valley effect phenomenon states that anthropomorphic appearance of a robot leads to trust and familiarity (i.e., humanoid robots are preferred more by ASD individuals than non-humanoids due to human-like appearance and interactions) (Arora et al., 2021; Sung et al., 2007; Turkle, 2017). In HRI context, robotic anthropomorphism means association of human-like characteristics in humanoid robots (e.g., facial features of robots like big eyes, smiling face, interactive voice, speech, hand, and body gestures integrated into robots like ASIMO, NAO, Kirobo Mini, Pepper, etc.). The ability to anthropomorphize robots is strongly linked to attributing human personality traits related to user's personality leading to robot likeability. Robotic intentionality (a.k.a., intentional stance or intentional mindset) is governed by the assumption that humans' state-of-mind associations, mental states, anthropomorphic beliefs, and desires result in (positive or negative) behaviors toward robots (Dennett, 1971, 1987).

A key hypothesis after this effort states that social robots may overcome some of the motivating and emotional challenges; they face with people with ASD when they interact with their partners (Dautenhahn, 1999). Contrary to their developing peers, whose interactions reward them naturally, children with ASD show only a weak function of the brain reward system in response to social reinforcement (Chevallier et al., 2012; Delmonte et al., 2012; Watson et al., 2015). ASD Community Outreach Vision Chevallier et al. (2012) stated that children with ASD do not want to maintain relationships with human partners, instead showing a preference for non-human and often mechanical objects (Watson et al., 2015).

In addition to these encouraging problems, neuropathy for people with ASD is uncommon: they often do not tolerate the complexities of many things (Bogdashina, 2010, 2012), show detailed data (Happé and Frith, 2006), and sensory sensitivity or conflict (Bogdashina, 2010), with significant social concerns (Spain et al., 2018). According to the theory of Weak Central Coherence (Happé et al., 2001) and the Enhanced Perceptual Functioning model (Mottron et al., 2006), cognitive processing of ASD individuals focuses on local structures. These children are unable to integrate individual pieces of patterns of the world. The Intense World Theory of Autism (Markram, 2007) suggested that these individuals suffer from excessive neuronal data processing resulting in detailed loading and abnormal levels of anxiety, which they seek to reduce with repeated and repetitive behaviors (Rodgers et al., 2012).

Over the past 50 years, there has been a growing interest in social cognition, which has prompted researchers to investigate new issues related to attributing ideas to others. Social cognition refers to the psychological operations that are the basis of social interaction, including perception, interpretation, and response to the intentions, personality, and behavior of others (Green, Horan, 2010). The ability to think and predict preferences, thoughts, desires, thinking, behavioral reactions, plans, and beliefs of others is an important aspect of social cognition (Frith, Frith, 2012) and is often referred to as "mindreading" or "mentalization."

Some of the most critical challenges people with ASD face are social interactions and their social and emotional development. This difficulty in communicating and interacting socially results from impaired language and communication skills, often combined with a lack of cognitive skills. Individuals with neurotypical development have communication skills based on their capacities for social interactions. However, it is difficult to focus on developing separated communication and entertainment. The term sociability refers to a person's ability to adapt to social situations and engage in friendly and professional relationships. Estimating the proper level of anthropomorphic robots used in the treatment of ASD is important. If the humanoid looks like a human, the child may begin to feel fear and apathy. On the other hand, it should not look like a machine because the child will be more interested in testing it than interacting with it. Humanoids adopt a beautiful, attractive design (i.e., large eyes, posture, body language, and facial expressions) to give a rich speech and help prevent fear in children with ASD.

Given the characteristics of ASD, it seems helpful to consider that a social robot with dynamic motivation, behavioral repetition, simplified appearance, and a lack of judgment in society may be more appealing to people with ASD than real people. Under the Intense World Theory of Autism (Markram, 2007), there is a reduction in unpleasant anxiety-related behaviors (e.g., superstitions, shouting, spontaneous attacks, etc.) during a human–robot interaction (HRI) situation. Therefore, in line with the ASD Social Motivation theory (Chevallier et al., 2012) and SCT theory, we propose the following propositions.

Proposition 1 Human-robot interaction (HRI) between (anthropomorphic and intentional) social robots and ASD individuals results in better social motivation and cognition for ASD individuals and individuals with other learning disorders / disabilities.

Proposition 2 HRI between (anthropomorphic and intentional) social robots and ASD individuals results in a reduction in unpleasant anxiety-related behaviors.

ROBOTIC ANTHROPOMORPHISM, INTENTIONALITY, SOCIAL COGNITION, AND AUTISM: DISCUSSIONS AND CONCLUSIONS

Anthropomorphism refers to giving personality traits or human-like characteristics to non-human objects, such as robots, computers, and animals. Anthropomorphizing objects is a way to build relationships with them, to deal with them as mediators in a communicative interaction. This process leads to the automatic delivery of intentionality and social behavior. Intentional stance allows us to deal with unknown entities or artifacts whose behavior is ambiguous or impregnable. Research in social robotics and HRI explores the impact of attributing deliberateness to mechanisms and the robot's behavioral parameters that almost all with efficiency induce this. In examining the effects that attributing deliberateness has on social interaction, participants of associate degree experiment are typically semiconductor diode to believe that they're interacting either with a pre-programmed machine (e.g., Wykowska et al., 2014; Özdem et al., 2017) or with another human (who naturally has wishes and beliefs). In other studies, participants are initially exposed to totally different agent sorts (e.g., human, mechanical man mechanism, non-humanoid robot) and afterward are semiconductor diode to believe that they're interacting with one among them (e.g., Krach et al., 2008). causation a specific (e.g., intentional) stance through instruction manipulation stands in distinction to analysis ways that aim to outline the parameters beneath that participant's ad libitum assume the intentional stance. Here, the intentional stance is induced through, as an example, the robot's gaze, speech, or general behavior (e.g., Wykowska et al., 2015). Godspeed questionnaires measure robotic anthropomorphism and intentionality. The full survey instruments are available as Appendix 5.1 and Appendix 5.2.

Social cognition is defined as understanding, perceiving, and interpreting information about other people and ourselves in a social context. These include emotional recognition, cognitive theory (ToM), delivery style, social vision, and knowledge. Social cognition consists of various processes that allow people to understand and interpret rapidly changing social data and respond appropriately to social incentives quickly, effortlessly, and easily. Recent works have shown that cognitive functioning and social skills among autistic individuals in proven steps only show a slight correlation in their functional outcomes over other factors (Sasson et al., 2020). Some people with autism may exhibit general social skills despite a low perception of psychological functioning with psychological compensation (Livingston et al., 2019). Among adults with autism without cognitive impairment, general cognition predicts more social potential than social cognition (Sasson et al., 2020), and the performance of explicit social-cognitive measures such as those used here may be less predictable social interaction Behavioral Behavior in Autism rather than Practice Social clarity (Keifer et al., 2020). Using natural methods is a challenge in terms of experimental control. Humanoid robots can prove particularly useful in this context, as they allow studying social cognition and joint attention specifically with a high degree of experimental control and relatively high ecological validity. That approach provides new insights into collaborative attention-based approaches (such as the role of human similarity, eye contact in visual acuity outcomes, difficulty severing facial expressions), and the ability to apply for health care, training, and assessment of joint care for children diagnosed with ASD. Appendix 5.3 provides the measures for social cognition targeted at ASD individuals during human–robot interaction (HRI) situations.

In conclusion, individuals with autism face behavioral challenges, and social robots can help to mitigate those uncommon behaviors / challenges. ASD individuals have difficulty communicating with other people—often failing to see people as human beings rather than simply being things in their environment. They cannot communicate easily with ideas and feelings, have difficulty concentrating on what others think or feel, and sometimes spend their lives in silence. They often find it challenging to make friends or even to bond with family members. Studying the conditions and consequences of implementing human-like behaviors on artificial agents that can potentially induce the adoption of an intentional stance is fascinating from a theoretical perspective and extremely important for the future of our societies.

APPENDICES

Appendix 5.1 Godspeed Questionnaires—Measures of Anthropomorphism (Adapted from Bartneck et al., 2009)

GODSPEED I: ANTHROPOMORPHISM

Please rate your impression of the robot on these scales:
以下のスケールに基づいてこのロボットの印象を評価してください。

Fake 偽物のような	1	2	3	4	5	Natural 自然な
Machinelike 機械的	1	2	3	4	5	Humanlike 人間的
Unconscious 意識を持たない	1	2	3	4	5	Conscious 意識を持っている
Artificial 人工的	1	2	3	4	5	Lifelike 生物的
Moving rigidly ぎこちない動き	1	2	3	4	5	Moving elegantly 洗練された動き

GODSPEED II: ANIMACY

Please rate your impression of the robot on these scales:
以下のスケールに基づいてこのロボットの印象を評価してください。

Dead 死んでいる	1	2	3	4	5	Alive 生きている
Stagnant 活気のない	1	2	3	4	5	Lively 生き生きとした
Mechanical 機械的な	1	2	3	4	5	Organic 有機的な
Artificial 人工的な	1	2	3	4	5	Lifelike 生物的な
Inert 不活発な	1	2	3	4	5	Interactive 対話的な
Apathetic 無関心な	1	2	3	4	5	Responsive 反応のある

GODSPEED III: LIKEABILITY

Please rate your impression of the robot on these scales:
以下のスケールに基づいてこのロボットの印象を評価してください。

Dislike 嫌い	1	2	3	4	5	Like 好き
Unfriendly 親しみにくい	1	2	3	4	5	Friendly 親しみやすい
Unkind 不親切な	1	2	3	4	5	Kind 親切な
Unpleasant 不愉快な	1	2	3	4	5	Pleasant 愉快な
Awful ひどい	1	2	3	4	5	Nice 良い

GODSPEED IV: PERCEIVED INTELLIGENCE

Please rate your impression of the robot on these scales:
以下のスケールに基づいてこのロボットの印象を評価してください。

Incompetent 無能な	1	2	3	4	5	Competent 有能な
Ignorant 無知な	1	2	3	4	5	Knowledgeable 物知りな
Irresponsible 無責任な	1	2	3	4	5	Responsible 責任のある
Unintelligent 知的でない,	1	2	3	4	5	Intelligent 知的な
Foolish 愚かな	1	2	3	4	5	Sensible 賢明な

GODSPEED V: PERCEIVED SAFETY

Please rate your emotional state on these scales:
以下のスケールに基づいてあなたの心の状態を評価してください。

Anxious 不安な	1	2	3	4	5	Relaxed 落ち着いた
Agitated 動揺している	1	2	3	4	5	Calm 冷静な
Quiescent 平穏な	1	2	3	4	5	Surprised 驚いた

Appendix 5.2 Measures of Intentionality and Negative Attitude Toward Robots (Adapted from Nomura et al., 2006)

Intentional Stance / Intentionality Questionnaire (ISQ, Marchesi et al., 2019) can be found at: https://instanceproject.eu/publications/rep
The complete Instance Questionnaire can be found at: https://drive.google.com/file/d/1DFY8lXB9uyR8LqPvxoQ-2hAVLXriWY5z/view

NEGATIVE ATTITUDE TOWARD ROBOTS SCALE

Appendix 5.3 Measures of Social Motivation Targeted at ASD individuals during Human-Robot Interaction (HRI) Situations (Adapted from Lang & Carstensen, 2002)

See Tables 5.1 and 5.2.

Table 5.1 The English version of negative attitude toward robots scale and the subscales that each item is included

Item no	Questionnaire items	Subscale
1	I would feel uneasy if robots really had emotions	S2
2	Something bad might happen if robots developed into living beings	S2
3	I would feel relaxed talking with robots*	S3
4	I would feel uneasy if I was given a job where I had to use robots	S1
5	If robots had emotions, I would be able to make friends with them*	S3
6	I feel comforted being with robots that have emotions*	S3
7	The word "robot" means nothing to me	S1
8	I would feel nervous operating a robot in front of other people	S1
9	I would hate the idea that robots or artificial intelligences were making judgments about things	S1
10	I would feel very nervous just standing in front of a robot	S1
11	I feel that if I depend on robots too much, something bad might happen	S2
12	I would feel paranoid talking with a robot	S1
13	I am concerned that robots would be a bad influence on children	S2
14	I feel that in the future society will be dominated by robots	S2

*Reverse item

Table 5.2 Social motivation questionnaire. Instruction: Read the following statements. For each statement, please judge how much do you agree with it according to your own situation. Shade the oval under the appropriate number on the scale, where 1 means "very disagree" and 7 means "very agree"

Item	1	2	3	4	5	6	7
1 It is important for me to spend time with people who know about topics that I know very little about	o	o	o	o	o	o	o
2 I seek contact with people who accept me the way I am	o	o	o	o	o	o	o
3 At this point in my life it is important for me to contact knowledgeable persons	o	o	o	o	o	o	o
4 I spend most of my time with people whom I feel very close	o	o	o	o	o	o	o
5 Few things are more interesting than meeting new and different people	o	o	o	o	o	o	o
6 I need to be with people who give my life a sense of meaning	o	o	o	o	o	o	o
7 I like to be with people who challenge my intellect	o	o	o	o	o	o	o
8 At my age there should always be someone around with whom there is a sense of mutual understanding	o	o	o	o	o	o	o

Note IS = information-seeking social motivation; ER = emotion-regulatory social motivation. The total score of item 1, 3, 5, and 7 measure the level of IS motivation, with a higher score indicating stronger IS motivation; the total score of item 2, 4, 6, and 8 measure the level of ER motivation, with a higher score indicating stronger ER motivation

REFERENCES

Aggarwal, P., & McGill, A. L. (2007). Is that car smiling at me? Schema congruity as a basis for evaluating anthropomorphized products. *Journal of Consumer Research, 34*(4), 468–479.

Arora, A. S., Fleming, M., Arora, A., Taras, V., & Xu, J. (2021). Finding "H" in HRI: Examining human personality traits, robotic anthropomorphism, and robot likeability in human-robot interaction. *International Journal of Intelligent Information Technologies (IJIIT), 17*(1), 19–38.

Arora, A. S., & Arora, A. (2020). The Race between Cognitive and Artificial Intelligence: Examining Socio-Ethical Collaborative Robots through Anthropomorphism and Xenocentrism in Human-Robot Interaction. *International Journal of Intelligent Information Technologies (IJIIT), 16*(1), 2020.

Bartneck, C., Kulić, D., Croft, E., & Zoghbi, S. (2009). Measurement instruments for the anthropomorphism, animacy, likeability, perceived intelligence, and perceived safety of robots. *International Journal of Social Robotics, 1*(1), 71–81.

Begum, M., Serna, R. W., & Yanco, H. A. (2016). Are robots ready to deliver autism interventions? A comprehensive review. *International Journal of Social Robotics*, *8*(2), 157–181.

Billard, A., Robins, B., Nadel, J., & Dautenhahn, K. (2007). Building robota, a mini-humanoid robot for the rehabilitation of children with autism. *Assistive Technology*, *19*, 37–49.

Boccanfuso, L., & O'Kane, J. M. (2011). Charlie: An adaptive robot design with hand and face tracking for use in autism therapy. *International Journal of Social Robotics*, *3*, 337–347.

Bogdashina, O. (2010). *Autism and the edges of the known world: Sensitivities, language, and constructed reality*. Jessica Kingsley.

Chevallier, C., Kohls, G., Troiani, V., Brodkin, E. S., & Schultz, R. T. (2012). The social motivation theory of autism. *Trends in Cognitive Sciences, 16*, 231–239. https://doi.org/10.1016/j.tics.2012.02.007

Clabaugh, C., Mahajan, K., Jain, S., Pakkar, R., Becerra, D., Shi, Z., et al. (2019). Long-term personalization of an in-home socially assistive robot for children with autism spectrum disorders. *Frontiers in Robotics and AI*, *6*, 110.

Damiano, L., & Dumouchel, P. (2018). Anthropomorphism in human-robot co-evolution. *Frontiers in Psychology, 9,*. https://doi.org/10.3389/fpsyg.2018.00468

Delmonte, S., Balsters, J. H., McGrath, J., Fitzgerald, J., Brennan, S., Fagan, A. J., & Gallagher, L. (2012). Social and monetary reward processing in autism spectrum disorders. *Molecular Autism, 3*(1), 1–13.

Dennett, D. C. (1971). Intentional systems. *The Journal of Philosophy, 68*, 87–106. https://doi.org/10.2307/2025382

Dennett, D. C. (1987). The Intentional Stance. MIT Press.

Diehl, J. J., Schmitt, L. M., Villano, M., & Crowell, C. R. (2012). The clinical use of robots for individuals with autism spectrum disorders: A critical review. *Research in Autism Spectrum Disorders, 6*(1), 249–262.

Emanuel, R., & Weir, S. (1976, July). Catalysing communication in an autistic child in a LOGO-like learning environment. In *Proceedings of the 2nd Summer Conference on Artificial Intelligence and Simulation of Behaviour* (pp. 118–129).

Francois, D., Stuart, P., & Dautenhahn, K. (2009). A long-term study of children with autism playing with a robotic pet: Taking inspirations from non-directive play therapy to encourage children's proactivity and initiative-taking. *Interaction Studies, 10*, 324–373.

Frith, C. D., & Frith, U. (2012). Mechanisms of social cognition. *Annual review of psychology, 63*(1), 287–313.

Ghorbandaei Pour, A., Taheri, A., Alemi, M., & Meghdari, A. (2018). Human-robot facial expression reciprocal interaction platform: Case studies on children with autism. *International Journal of Social Robotics, 10*, 179–198.

Giannopulu, I. (2013). Multimodal cognitive nonverbal and verbal interactions: The neurorehabilitation of autistic children via mobile toy robots. *International Journal of Advances in Life Sciences, 5,* 214–222.

Golestan, S., Soleiman, P., & Moradi, H. (2017). Feasibility of using sphero in rehabilitation of children with autism in social and communication skills. In *2017 International Conference on Rehabilitation Robotics (ICORR)* (pp. 989–994). IEEE.

Green, M. F., & Horan, W. P. (2010). Social cognition in schizophrenia. *Current Directions in Psychological Science, 19*(4), 243–248.

Guizzo, E. (2015). Jibo is as good as social robots get. But is that good enough? *Science Robotics, 3,* 21.

Happé, F., & Frith, U. (2006). The weak coherence account: Detail-focused cognitive style in autism spectrum disorders. *Journal of Autism and Developmental Disorders, 36*(1), 5–25.

Happé, F., Frith, U., & Briskman, J. (2001). Exploring the cognitive phenotype of autism: weak "central coherence" in parents and siblings of children with autism: I. Experimental tests. *The Journal of Child Psychology and Psychiatry and Allied Disciplines, 42*(3), 299–307.

Howe, J. (1983). Autism-using a 'turtle' to establish communication. In W. J. Perkins (Ed.), *High Technology Aids for the Disabled,* pp. 179–183. Elsevier. https://doi.org/10.1016/B978-0-407-00256-2.50033-2

Ismail, L. I., Verhoeven, T., Dambre, J., & Wyffels, F. (2019). Leveraging robotics research for children with autism: A review. *International Journal of Social Robotics, 11*(3), 389–410.

Jobe, L. E., & White, S. W. (2007). Loneliness, social relationships, and a broader autism phenotype in college students. *Personality and Individual Differences, 42*(8), 1479–1489.

Keifer, C. M., Mikami, A. Y., Morris, J. P., Libsack, E. J., & Lerner, M. D. (2020). Prediction of social behavior in autism spectrum disorders: Explicit versus implicit social cognition. *Autism, 24,* 1758–1772. https://doi.org/10.1177/1362361320922058

Kim, E. S., Berkovits, L. D., Bernier, E. P., Leyzberg, D., Shic, F., Paul, R., et al. (2013). Social robots as embedded reinforcers of social behavior in children with autism. *Journal of Autism and Developmental Disorders, 43,* 1038–1049.

Krach, S., Hegel, F., Wrede, B., Sagerer, G., Binkofski, F., & Kircher, T. (2008). Can machines think? Interaction and perspective taking with robots investigated via fMRI. *PloS One, 3*(7): e2597.

Lang, F. R., & Carstensen, L. L. (2002). Time counts: Future time perspective, goals, and social relationships. *Psychology and Aging, 17*(1), 125.

Lebersfeld, J. B., Brasher, C., Biasini, F., & Hopkins, M. (2019). Characteristics associated with improvement following the SAM robot intervention for children with autism spectrum disorder. *International Journal Pediatric Neonatal Care*, 5, 9.

Livingston, L. A., Colvert, E., Bolton, P., & Happé, F. (2019). Good social skills despite poor theory of mind: Exploring compensation in autism spectrum disorder. *Journal of Child Psychology and Psychiatry*, 60, 102–110. https://doi.org/10.1111/jcpp.12886

Lord, C., Rutter, M., & Le Couteur, A. (1994). Autism Diagnostic Interview-Revised: A revised version of a diagnostic interview for caregivers of individuals with possible pervasive developmental disorders. *Journal of Autism and Developmental Disorders*, 24(5), 659–685.

MacNeil, B. M., Lopes, V. A., & Minnes, P. M. (2009). Anxiety in children and adolescents with autism spectrum disorders. *Research in Autism Spectrum Disorders*, 3(1), 1–21.

Marchesi, S., Ghiglino, D., Ciardo, F., Perez-Osorio, J., Baykara, E., & Wykowska, A. (2019). Do we adopt the intentional stance toward humanoid robots? *Frontiers in Psychology*, 10, 450.

Markram, H. (2007). The intense world syndrome-an alternative hypothesis for autism. *Frontiers in Neuroscience*, 1, 77–96. https://doi.org/10.3389/neuro.01.1.1.006.2007

Mazurek, M. L. (2014). *A Tag-Based, Logical Access-Control Framework for Personal File Sharing*. Doctoral dissertation, Carnegie Mellon University.

Miyamoto, E., Lee, M., Fujii, H., & Okada, M. (2005). How can robots facilitate social interaction of children with autism? Possible implications for educational environments. In *Proceedings of the 5th International Workshop on Epigenetic Robotics: Modeling Cognitive Development in Robotic Systems*, 145–146.

Mottron, L., Dawson, M., Soulières, I., Hubert, B., & Burack, J. (2006). Enhanced perceptual functioning in autism: An update, and eight principles of autistic perception. *Journal of Autism and Developmental Disorders*, 36(1), 27–43.

Nomura, T., Suzuki, T., Kanda, T., & Kato, K. (2006, July). Altered attitudes of people toward robots: Investigation through the Negative Attitudes toward Robots Scale. In *Proc. AAAI-06 workshop on human implications of human-robot interaction* (Vol. 2006, pp. 29–35).

Özdem Yilmaz, Y., Cakiroglu, J., Ertepinar, H., & Erduran, S. (2017). The pedagogy of argumentation in science education: science teachers' instructional practices. *International Journal of Science Education*, 39(11), 1443–1464.

Pfeiffer, U. J., Timmermans, B., Bente, G., Vogeley, K., & Schilbach, L. (2011). A non-verbal turing test: Differentiating mind from machine in gaze-based social interaction. PLoS ONE 6:e27591. https://doi.org/10.1371/journal.pone.0027591

Pioggia, G., Sica, M. L., Ferro, M., Igliozzi, R., Muratori, F., Ahluwalia, A., et al. (2007). Human-robot interaction in autism: FACE, an android-based social therapy. In *RO-MAN 2007 - The 16th IEEE international symposium on robot and human interactive communication* (pp. 605–612). IEEE.

Rodgers, J., Glod, M., Connolly, B., & McConachie, H. (2012). The relationship between anxiety and repetitive behaviours in autism spectrum disorder. *Journal of Autism and Developmental Disorders, 42*, 2404–2409. https://doi.org/10.1007/s10803-012-1531-y

Rutter, M., Anthony, B., & Lord, C. (2003). *The Social Communication Questionnaire Manual.* Western Psychological Services.

Sasson, N. J., Morrison, K. E., Kelsven, S., & Pinkham, A. E. (2020). Social cognition as a predictor of functional and social skills in autistic adults without intellectual disability. *Autism Research, 13*, 259–270. https://doi.org/10.1002/aur.2195

Shamsuddin, S., Yussof, H., Ismail, L. I., Mohamed, S., Hanapiah, F. A., & Zahari, N. I. (2012). Initial response in HRI- a case study on evaluation of child with autism spectrum disorders interacting with a humanoid robot NAO. *Procedia Engineering, 41*, 1448–1455.

She, T., Kang, X., Nishide, S., & Ren, F. (2018). Improving LEO robot conversational ability via deep learning algorithms for children with autism. In *2018 5th IEEE International Conference on Cloud Computing and Intelligence Systems (CCIS)* (pp. 416–420). IEEE.

Spain, D., Sin, J., Linder, K. B., McMahon, J., & Happé, F. (2018). Social anxiety in autism spectrum disorder: A systematic review. *Research in Autism Spectrum Disorders, 52*, 51–68.

Sung, J.-Y., Guo, L., Grinter, R. E., & Christensen, H. I. (2007). *My roomba is ramboi: Intimate home appliances.* Springer.

Turkle, S. (2017). Alone together: Why we expect more from technology and less from each other.

Terada, K., Shamoto, T., Ito, A., & Mei, H. (2007, October). Reactive movements of non-humanoid robots cause intention attribution in humans. In *2007 IEEE/RSJ International Conference on Intelligent Robots and Systems* (pp. 3715–3720). IEEE.

Terada, K., Shamoto, T., & Ito, A. (2008, August). Human goal attribution toward behavior of artifacts. In *RO-MAN 2008-The 17th IEEE International Symposium on Robot and Human Interactive Communication* (pp. 160–165). IEEE.

Valadao, C. T., Goulart, C., Rivera, H., Caldeira, E., Bastos Filho, T. F., Frizera-Neto, A., et al. (2016). Analysis of the use of a robot to improve social skills in children with autism spectrum disorder. *Research on Biomedical Engineering, 32*, 161–175.

Vickerstaff, S., Heriot, S., Wong, M., Lopes, A., & Dossetor, D. (2007). Intellectual ability, self-perceived social competence, and depressive symptomatology in children with high-functioning autistic spectrum disorders. *Journal of Autism and Developmental Disorders*, *37*(9), 1647–1664.

Wainer, J., Dautenhahn, K., Robins, B., & Amirabdollahian, F. (2014). A pilot study with a novel setup for collaborative play of the humanoid robot KASPAR with children with autism. *International Journal of Social Robotics*, *6*, 45–65.

Watson, K. K., Miller, S., Hannah, E., Kovac, M., Damiano, C. R., Sabatino-DiCrisco, A., et al. (2015). Increased reward value of non-social stimuli in children and adolescents with autism. *Frontiers in Psychology*, *6*, 1026. https://doi.org/10.3389/fpsyg.2015.01026

Werry, I. P., & Dautenhahn, K. (1999). Applying mobile robot technology to the rehabilitation of autistic children. In *Proceedings of SIRS99, 7th Symp on Intelligent Robotic Systems*, 265–272.

Werry, I., Dautenhahn, K., Ogden, B., & Harwin, W. (2001). Can social interaction skills be taught by a social agent? The role of a robotic mediator in autism therapy. In *CT'01 Proceedings of the 4th International Conference on Cognitive Technology: Instruments of Mind*. Springer-Verlag. https://doi.org/10.1007/3-540-44617-6_6

Wykowska, A., Kajopoulos, J., Obando-Leiton, M., Chauhan, S. S., Cabibihan, J. J., & Cheng, G. (2015). Humans are well tuned to detecting agents among non-agents: examining the sensitivity of human perception to behavioral characteristics of intentional systems. *International Journal of Social Robotics*, *7*(5), 767–781.

Wykowska, A., Wiese, E., Prosser, A., & Müller, H. J. (2014). Beliefs about the minds of others influence how we process sensory information. *PloS One*, *9*(4), e94339.

Yamaji, Y., Miyake, T., Yoshiike, Y., De Silva, P. R., & Okada, M. (2010, March). STB: Human-dependent sociable trash box. In *2010 5th ACM/IEEE International Conference on Human-Robot Interaction (HRI)* (pp. 197–198). IEEE.

Yang, Y., Li, Y., Fermuller, C., & Aloimonos, Y. (2015, March). Robot learning manipulation action plans by "watching" unconstrained videos from the World Wide Web. In *Proceedings of the AAAI conference on artificial intelligence* (Vol. 29, No. 1).

Yoshikawa, Y., Kumazaki, H., Matsumoto, Y., Miyao, M., Kikuchi, M., & Ishiguro, H. (2019). Relaxing gaze aversion of adolescents with autism spectrum disorder in consecutive conversations with human and android robot-a preliminary study. *Frontiers in Psychiatry*, *10*, 370.

Digital Technology

Artificial Intelligence (AI) in Marketing: How AI Supports Marketers Throughout the Consumer Journey

Emily Osbourne and Anshu Saxena Arora

Abstract With an increase in the volume of consumer-curated data, marketers, and advertisers rely increasingly on Artificial Intelligence (AI) to transform this data into valuable consumer insights. AI's key building blocks, such as Natural Language Process (NLP), image recognition, speech recognition, problem solving and reasoning, and machine learning, provide a gateway for marketers to understand and predict the needs and demands of consumers as well as reach them at all levels of their consumer journey. The purpose of this research is to understand deep linkages and ties between AI applications and subsequent AI relationships with consumers through digital marketing interfaces of various

E. Osbourne · A. S. Arora (✉)
University of the District of Columbia, Washington, DC, USA
e-mail: anshu.arora@udc.edu

E. Osbourne
e-mail: emily.osbourne@udc.edu

A. S. Arora et al. (eds.), *Managing Social Robotics and Socio-cultural Business Norms*, International Marketing and Management Research, https://doi.org/10.1007/978-3-031-04867-8_6

organizations. Three case studies are presented in this chapter to examine the proposed framework through AI Applications—Relationships lenses and implications for researchers, marketers, and policy makers are offered.

Keywords Data · Artificial intelligence · Consumer journey · Building blocks · Unstructured data · Structured data

INTRODUCTION

Artificial Intelligence (AI) has recently become a very popular subject in the field of management and marketing due to its advancements and potential applicability in influencing consumer decision-making behavior. As the market continues to expand, consumers are now receiving countless offers to shop, which has subsequently created a seemingly endless supply of consumer curated data that expresses their desires, attitudes, and beliefs (Kietzman et al., 2018). Due to this exponential increase of data and continual rise of customer expectations, marketers are now beginning to rely on AI to better track and understand consumer behavior at each stage of the consumer journey. This is because AI is data-driven in nature and can transform the data received into useful strategies that can then be used to guide consumer behavior, leading to higher rates of customer satisfaction (Rabby & Hassan, 2021).

Artificial intelligence (AI) is defined as "mimicking human cognition by computers / robots (Jha and Topol, 2016).... with a focus on human intelligence, human thinking, and problem solving skills since its inception in 1956 (Dautenhahn, 2007)" (Arora & Arora, 2020, p. 3; Arora et al., 2021). In other words, AI is a constellation of technologies—from machine learning to natural language processing—that allows machines to sense, comprehend, act, and learn (Batra, 2019). AI leverages its self-learning systems and building blocks to minimize errors, make continuous improvements, and translate data into valuable insight. These building blocks include natural language processing (NLP), image recognition, speech recognition, problem solving and reasoning, as well as, machine learning (ML). NLP is a system that allows the AI to analyze the nuances of human language and create meaning from posts and blogs, among other channels used by consumers to express their opinions. Image recognition is another tool used by advertisers to understand the content that is shared and posted on social media, revealing a consumer's "real" behavior. Speech recognition is an analysis of the words spoken, while problem solving, and reasoning is used to detect data patterns and make future predictions on consumer preferences based on pre-existing data. Lastly, machine learning (ML), like problem solving, works by proposing the best option for consumer needs, predicts customer lifetime value and conversion likelihood as it accumulates data overtime (Kietzman et al., 2018). These systems work together to create a more meaning online experience for consumers, as the information displayed is tailored toward their specific tastes.

Further, it is important to note the consumer purchase journey. This begins with the need and/or want recognition where consumers realize their need for a specific product or service is triggered, followed by initial consideration of possible offerings leading to active evaluation and customer(s) narrowing down brand choices, followed by subsequent brand purchase (transaction is made as consumers determine how much they are willing to pay), and post purchase (Kietzman et al., 2018), whereby overall customer satisfaction / dissonance is evaluated. AI is used to influence customers at every stage of this journey using the building blocks previously described. This can be seen in the use of features such as chatbots that allow customers to speak with virtual assistants during shopping, thus decreasing their waiting time, as well as, the pressure on employees (Abu Daqar & Smoudy, 2019).

Despite the array of information which now exists on AI in marketing, past research has illustrated gaps in describing the role of AI in marketing, specifically its impact on customer experiences through customer service and after-sale customer support. Abu Daqar and Smoudy (2019) highlights the benefits, firm(s) receive from having satisfied customers, however, they didn't provide details regarding the process of receiving and analyzing consumer data to create personalized consumer recommendations. Further, previous research illustrates the business implications of implementing AI into marketing, designing innovations and provides ideas on how to seamlessly incorporate these new skills into marketing teams (Batra, 2019; Jarek & Mazurek, 2019). AI research, however, did not mention what this implementation of new technology means for individuals working within these areas that are now becoming automated. Additionally, AI's influence on buyer behavior and marketing decisions is explored through an investigation into AI's relationships with improving service quality and consumer trust in digital platforms (Abu Daqar &

Smoudy, 2019). Current and past AI research have yet to examine the dangers of AI in relation to privacy concerns of the consumers/users in digital and traditional world. As such, in an attempt to bridge these gaps, this research addresses the following questions:

1. How does AI work in global business transactions today?
2. What are the implications of AI on jobs and society?
3. What dangers / issues exist for consumers in terms of privacy with AI in marketing?

This research seeks to add value to industry, specifically the areas of management and marketing, as AI proves to be revolutionary in achieving customer satisfaction and competitive advantage within the industry. By addressing the above questions, we provide consumers, managers, and markets better insights into the inner workings of AI and its impact on society. The article also addresses the privacy concerns of consumers and businesses that may have been exemplified through the utilization and implementation of AI.

THEORETICAL BACKGROUND

Artificial Intelligence in Business Applications

As previously stated, Artificial Intelligence or AI learns from revising and monitoring both the past and current actions of consumers using its key building blocks. When someone uses their electronic devices to browse the Internet, their data are collected and manipulated in real time though computer programs that create and send recommendations based on the consumers' interests and behaviors expressed through likes, clicks, and searches (Murgai, 2018). This mimics a customer retention process since it creates a personalized experience for its users. AI based applications or technologies are used by customers and may take different forms, including (but not limited to) email spam filters, predictive search terms, virtual assistants such as Siri and Alexa, as well as, recommended news items and shopping brands (Batra, 2019). Other recent AI developments are:

- Search option results: AI is always seeking to strengthen its prediction of a consumer's wants or needs before they express it through the common search option usually located on a company's website or app. This behavioral information is now becoming more available to brands, marketers, and corporations alike to help improve and streamline user experiences and their consumer (purchase) journey by way of offering tailored recommendations. One such example of this is Google, which uses machine learning to improve a consumer's algorithm with every search executed (Batra, 2019).

- "ALWAYS ON" is another AI application which collected data on consumer insights non-stop, behind-the-scenes through the passive user interface between the smart devices and cloud data. The passive insights are used to gain an additional understanding of consumer behavior through machine learning (Batra, 2019).

Artificial Intelligence: Employment and Society

With the increasing role of AI in marketing, it is only natural to question what this means for the labor market. While AI and automation can improve productivity and reduce capital costs, it can also replace the work being done by others and transform many occupations in the process (Ernst et al., 2019). In general, depending on the nature of the job, a worker may be augmented by technology or be in competition with it (Frank et al., 2019). AI has proven to be faster and more accurate in anticipating the desires of consumers and executing them, leaving more time for workers to focus on other tasks that might be too complex for the application.

During the automation process of firms, job growth tends to be affected through three channels. First, new technologies lead to a substitution of jobs and tasks being performed by workers (the displacement effect). When some tasks become modified due to AI, job profiles might change by adding or modifying existing tasks instead of suppressing the job entirely. This, however, depends on whether it remains profitable to group certain tasks that have not yet been automated with those that did not exist before (Ernst et al., 2019). Second, there is a complementary increase in jobs and tasks required to use, run, and supervise the new machines (the skill complementarity effect) (Petropoulos, 2018); and third, there is a demand effect from both the lower prices and a general increase in disposable income in the economy due to higher productivity (the productivity effect) (Ernst et al., 2019). The standard narrative in this concept is that unemployment will initially increase with the introduction to automation before falling again, when both prices and productivity adjust across the economy. It is also important to emphasise that when AI takes over repetitive or dangerous tasks, it frees up the human workforce to do work they are better equipped for—tasks that involve creativity, intuition and empathy among others. If people are doing work that is more engaging for them, it could increase happiness and job satisfaction (Marr, 2021). This implementation of AI coupled with the new responsibility of workers may lead to new opportunities for growth and development.

Privacy Issues with AI in Marketing

As a result of the large quantities of data now available, both structured and unstructured, AI and big data are now reshaping the risk in consumer privacy and data security. Naturally, this raises the concern of consumers on the issues related to AI. When used correctly by legitimate companies, AI could improve management efficiency, motivate innovations, and better match demand and supply. Alternatively, in the wrong hands, the information gained through AI can be used to conduct fraud, identity theft, and blackmail (Zhe Jin, 2018).

Typically, data can be stored, traded, and used long after a transaction has occurred. Some sellers may be reluctant to restrict data use to a particular purpose or data processing method; even if they do not intend to use the data from the transaction, they might be willing to sell it to those who will. These types of data markets motivate sellers to collect as much information as they can from customers (Zhe Jin, 2018). As such, buyers may experience trust issues due to AI innovation in digital marketing and, as a result, clients are keen to spend more time educating themselves before they commit to a purchase or give away personal data (Rabby & Hassan, 2021). However, in many transactions, identity and payment information are needed to complete sale (Zhe Jin, 2018). The damage from future misuse often affects only the consumer as opposed to the seller because it is often difficult to trace back consumer harm to a particular data collector, especially if the victim has shared the same information with hundreds of sellers and have no control over what information is given away. That said, businesses that are attempting to implement high quality AI must do so seamlessly to prevent any negative consequences for both itself and fellow consumers. To do so effectively, marketers and advertisers

Fig. 6.1 Relationship between AI in digital marketing and consumer behavior (Rabby & Hassan, 2021)

must understand the perception consumers have of AI as it could greatly influence the accuracy of results and the attractiveness of brands. As such, trust and confidence in AI are necessary to bridge the gap between quality and convenience, and AI applications support companies in taking advantage of technological developments as seen in Fig. 6.1 (Rabby & Hassan, 2021).

Conceptual Framework

The AI Applications—Relationships framework displayed in Fig. 6.2 showcases the various benefits of Artificial Intelligence in marketing. Through its use, AI can appeal to a person's needs and wants by providing a thorough understanding of the customer and their preferences based on past activities, which have been used to improve customer service. AI can also be used to create employment opportunities, as stated earlier,

Fig. 6.2 AI applications—relationships framework

with the implementation of new technologies, many workers now have the time to focus on other tasks that may require more creativity and empathy. Workers work directly with AI to better understand its workings and purpose, as well as, the value it seeks to add to the society. Lastly, through its improvements in customer service experiences, consumers will be able to have confidence in AI and trust that their information is being protected even after a transaction has occurred. These factors all contribute to influencing consumer behavior and improving / building consumer relationships with AI.

Artificial Intelligence has the potential to revolutionize the ways in which businesses interact with their customers. Today, its technologies are often embraced by firms to remain competitive, as well as, to respond to the sustained margin pressures, shorter strategy cycles and increased customer expectations (Ameen et al., 2020). These alterations have allowed companies to achieve better customer brand relationships as AI can significantly improve the customer experiences by increasing a company's knowledge about their preferences and shopping patterns through its building blocks. When deployed strategically at various key customer touch points, AI technologies can greatly increase brand relation and customer satisfaction through its personalized services and product recommendations based on a customer's past purchase behavior and preferences. In fact, reports conducted from earlier studies reported that AI can reach the top 1% of customers, who are worth 18 times more than average customers to retailers (Ameen et al., 2020). These results were achieved through increased personalization and engagement based on behavioral and contextual data.

- **Proposition 1:** *The use of AI will increase consumer satisfaction at each stage of the consumer (purchase) journey.*

Business CEOs and leaders have expressed their concerns with AI. Google's CEO, Sundar Pichai had said: "It is important to understand that tomorrow, whether Google is there or not, artificial intelligence is going to progress. Technology has this nature. It is going to evolve … AI is more important than fire or electricity" (Arora & Brown-Gaston, 2021). As Artificial Intelligence continues to become more sophisticated, it is inevitable that job displacement will occur. With AI and machine learning systems taking over repetitive tasks, and their capacity to process and transform data into valuable insight at a rapid rate, this creates a space for the emergence of new employment opportunities. These news roles, however, are not designed to replace the old but will require (new) appropriate training and skills. Based on this fact, researchers have revealed three new categories that can be foreseen with the implementation of AI driven technologies. This includes trainers, explainers, and sustainers (Wilson et al., 2017). **Trainers** will be needed to teach AI systems—how to perform and mimic human behaviors. An example of this are chatbots, they need to be trained to detect the complexities and subtleties of human communication. **Explainers** are the needed to bridge gaps between technology and business leaders, providing clarity and assisting in the analyzing of the results gathered from AI. **Sustainers** help to ensure that AI systems operate as designed and resolve fundamental issues. The role of human workers in these jobs complements the tasks being performed by AI (Wilson et al., 2017).

- ***Proposition 2:** The effective implementation of AI will lead to higher levels of production and employment within businesses.*

Customers often experience issues of trust with AI use and implementation as their information is shared across digital platforms by marketers and advertisers. This may be because of technophobia, the fear of advanced or complex technologies, issues of identity theft or fraud, as well as, ambiguity on the systems' operations. These factors can greatly determine whether a consumer is willing to trust AI or not. As such, an expert system needs to be able to justify and provide an explanation to users for its decisions; and information security should likewise be able to explain why a system is deemed secure (Pieter, 2010). Furthermore, companies should further implement customer data security measures and openly state the procedures taken to protect AI systems against potential intruders. With this information along with their understanding of the benefits behind the system, consumers will be able to successfully develop trust and self-assurance by assessment of risks and alternatives (Pieter, 2010).

- **Proposition 3:** *Consumers will trust and build confidence in AI when they experience benefits in AI usage and trust that companies (businesses) are implementing AI with appropriate data security and safety measures.*

CASE STUDIES

In this section, we explore our multi-dimensional, **AI Applications— Relationships framework** through the lenses of three real-life companies with AI applications in their digital marketing ecosystems. To conduct the research, three case studies were engaged, which is generally considered more robust than single case studies (Yin, 1994). The cases were deliberately selected and the case methodology presented here is consistent with the objectives of qualitative research (Glaser & Strauss, 1965; Silverman, 2000). The research methodology follows closely to qualitative works including Karjalainen and Snelders (2010), Brockman et al. (2010), and Mabert et al. (1992) that utilize case research to drive new framework or theory. We examine the consumer decision-making processes, AI-consumer interactions, and AI influences and applications in businesses. The cases, KLM , eBay , and 1-800-Flowers were purposely selected in consistency with the objectives of this research and our proposed framework. We explore our framework through these three cases and prove our propositions.

KLM

KLM, is an Amsterdam-based Royal Dutch Airline founded in 1919 and is the oldest global airline still flying under its original name. The company currently serves over 30 million passengers yearly and operates more than 200 aircrafts to its 163 destinations worldwide. The company has always been known for its proactive approach in developing technology and implementing AI to meet the rapid changing customer expectations.

On average, customers fly with KLM airlines approximately 1.4 times per year (KLM). Many of these passengers may not wish to download the airline's mobile app or manually book a flight. As such in September 2017, the KLM introduced a booking bot. The bot's personality was designed to be helpful, friendly, and professional. This was aimed toward helping customers book their tickets, send confirmation, delivery flight updates and answer passenger questions. With this type of AI implementation, customers can easily book a flight without having to rely on a KLM agent (KLM).

The systems find its footing in AI by connecting through KLM technology and is supported by 250 human service colleagues. The BlueBot (BB) is one of the most comprehensive customer support technologies, offered in up to 10 different languages and is trained on over 60,000 questions and answers, available 24 h a day. This allows the airline to respond to as many as 1.4 customer queries within a year (Daly, 2018).[1] If the BB is unable to answer a specific query, customers will automatically be connected to one of the airline's customer service agents. Since its implementation, KLM has reported a large increase in sales volume as customers were now able to get answers and assistance in real time, with accuracy. Without BlueBot, KLM could not be able to record more than 1.7 million messages being sent by their passengers (Mastorakis, 2018). The company also learned that customers love chatting with the BB as they can discover its different responses all kinds of questions.

The company is also planning on increasing the Bot's personalization of conversations. This is necessary as every conversation is unique and a response for one customer might not be the same for another. Many consumers may ask different variations in questions. BlueBot (BB) is now seeking to become more adaptable using the airline's historical case data (Daly, 2018).

eBay

eBay, one of the world's largest eCommerce platforms, connects millions of buyers and sellers together in more than 190 markets across the world. Focused on global trade, the issue of language barriers quickly became a challenge for many consumers using the website. This was because international buyers and sellers wanted to communicate but found it difficult to follow through with their business transactions.

eBay soon developed a machine learning system for language translation. This was aimed toward facilitating the transaction between users. The system is known as the eBay Machine Translation (eMT) (Langfelder,

---

[1] https://news.klm.com/klm-welcomes-bluebot-bb-to-its-service-family/.

2020).[2] During the early stages of its implementation, the application's use was mainly focused on Spanish and English translations on product tiles. eBay hoped that this would boost transaction activities between the United States and Latin America. It soon expanded its operations having received a 90% accuracy rate and a trade increase of 1.06% for each additional work in the title of items on eBay (Langfelder, 2020).

With these results, eBay further continued their use of eMT for all their translating needs. This included production descriptions and reviews. This AI and ML application is now used in many regions around the world to enable users to connect with each other despite their cultural and language differences (Langfelder, 2020) thus enabling eBay's global trade and international business.

1-800-Flowers

1–800-Flowers began as a small flower shop in Manhattan. Today, it has grown into one of the largest e-commerce retailers in the United States, making over 1.2 billion dollars in sales yearly. Despite still holding physical stores, the company conducts 88% of its operations online.

1-800-Flowers also has numerous acquisitions, which include The Popcorn Factory, Fannie May, and Cheryl's Cookies, all of which are integrated into their website. With its multiple sites and banners under which it does business, the process of navigating through its thousands of products can become confusing and frustrating for consumers. To resolve the issue, the company created GWYN.

GWYN, an acronym meaning "Gifts When You Need," is an AI-assistant powered by IBM that aims to transform a customer's online experience. The system gathers information on the gift recipient and interacts with customers through its cognitive capabilities to provide tailored product recommendations for any specific event needed by shoppers.[3] GWYN is essentially a high powered chatbot, accessible through mobile devices and desktop computers aimed at facilitating the consumer purchase journey. The system is trained to answer different types of questions, from vague to specific, and asks several follow-up questions to

---

[2] https://www.softworksai.com/machine-translation-and-ebays-attempted-comeback/.

[3] https://etaileast.wbresearch.com/blog/1-800-flowers-customer-journey-strategy-with-ai.

ensure the right product suggestions are provided to users. These includes questions regarding the occasion, sentiments, and likes and interests of the recipient(s). The more shoppers interact, the better intuitive GWYN becomes, allowing for a refined shopping experience over time.

GWYN, the bot, was an immediate success after its implementation. It led to a 70% increase in customer orders with most being from the younger generation (Langfelder, 2020). The AI application has also been efficient in attracting new shoppers to the company.[4]

DISCUSSIONS AND IMPLICATIONS

Artificial Intelligence has become an integral component for businesses seeking to remain competitive in today's industries. When used effectively, AI can not only assist, but also, boost customer satisfaction at the different stages of their consumer journey. This was evident in the case studies discussed in the previous section.

Proposition 1 (presented in our research) suggested that the use of AI will increase consumer satisfaction at each stage of the consumer journey. Using systems such as Chatbots, companies like KLM and 1-800-Flowers were able to claim an increase in sales, as well as, boost brand loyalty. In the case of KLM, the implementation of its BlueBot (BB) allowed customers to get assistance in real time, without having to wait for an agent of the airline, thus saving time. Similarly, 1800-Flowers' AI implementation of GWYN allowed customers to gain trust and become brand loyal based on the system's ability to create personalized product recommendations based on their tastes and preferences.

Proposition 2 (presented in our research) examines the implementation of AI that has the potential to increase levels of productivity and employment. For this proposition, all three case studies are cases in point. As previously stated, based on our AI Applications—Relationships framework created, the technologies formed by AI are not designed to replace existing tasks, but to complement the workings of system. With the implementation of these AI systems by KLM, eBay and 1800-Flowers, employment opportunities are created because skilled persons are now needed to teach AI how to perform and mimic the behavior of humans, assist in translating the results of AI to business leaders so they can make

---

[4] https://digital.hbs.edu/platform-rctom/submission/1-800-flowers-and-ibm-watson-take-on-the-future-of-gifting-and-the-future-of-relationships/.

strategic decisions, and sustain the technology and ensure that it operates smoothly without error. From the case studies, it is also evident that these companies were able to see an increase in productivity.

According to proposition 3 in our research, we propose that consumers will begin to trust and build confidence in AI, having seen the benefits of AI and the implementation of safety measures. For example, eBay, where buyers and sellers can participate in transactions across different borders, eBay has privacy policies located throughout its platform to inform customers of how the company is working to protect their personal data. This includes allowing only limited access to other user's contact, shipping, and financial information. This information, being shared by eBay's eMT, is only what is needed to successful complete the transaction and collect payments. Knowing that their information is being protected while they shop, gives consumers the comfort to shop online without having to worry about the misuse of their information.

AI has truly changed the way companies conduct their business. It has allowed marketers to be able to gain valuable insights on consumer interests based on their past behaviors, which can then be used to create customized recommendations for products and services consumers are most likely to enjoy. AI has also changed the employment scene, creating more jobs and opportunities for those individuals who are more inclined with technology. These people are needed at every stage of its implementation to ensure that the system can operate effectively. With the large number of online platforms that now exists and continuous AI implementation, many businesses are now advancing their security features and policies to ensure that the personal data collected from customers remain protected against issues such as fraud and identity theft. The future is to ensure that businesses are able to implement AI securely, efficiently and effectively so that the consumers can regain their trust and confidence in these AI applications.

REFERENCES

Abu Daqar, M. A., & Smoudy, A. K. (2019). The role of artificial intelligence on enhancing customer experience. *International Review of Management and Marketing, 9*(4), 22–31. https://doi.org/10.32479/irmm.8166

Arora, A. S., & Arora, A. (2020). The race between cognitive and artificial intelligence: Examining socio-ethical collaborative robots through anthropomorphism and xenocentrism in HRI. *International Journal of Intelligent Information Technologies, 16*(1), 1–16.

Arora, A. S., & Brown-Gaston, R. (2021). War and peace: Ethical challenges and risks in military robotics. *International Journal of Intelligent Information Technologies, 17*(3), 1–12.

Arora, A. S., Fleming, M., Arora, A., Taras, V., & Xu, J. (2021). Finding "H" in HRI: Examining human personality traits, robotic anthropomorphism, and robot likeability in human-robot interaction. *International Journal of Intelligent Information Technologies, 17*(1), 19–38.

Ameen, N., Tarhini, A., Reppel, A., et al. (2020). Customer experiences in the age of artificial intelligence. *Computers in Human Behavior, 114*(2021). https://doi.org/10.1016/j.chb.2020.106548

Batra, M. M. (2019). Strengthening customer experience through artificial intelligence: An upcoming trend. *Competition Forum, 17*(2), 223–231. Retrieved from https://www.proquest.com/scholarly-journals/strengthening-customer-experience-through/docview/2343014949/se-2?accountid=28903

Brockman, B. K., Rawlston, M. E., Jones, M. A., & Halstead, D. (2010). An exploratory model of interpersonal cohesiveness in new product development teams. *Journal of Product Innovation Management, 27*, 201–219.

Daly. C. (2018). KLM: Chatbots are the future of customer support. *AI Business.* Retrieved from: https://aibusiness.com/document.asp?doc_id=760517

Ernst, E., Merola, R., & Samaan, D. (2019). Economics of artificial intelligence: Implications for the future of work. *IZA Journal of Labor Policy, 9*(1). https://doi.org/10.2478/izajolp-2019-0004

Frank, M., Autor, D., Bessen, J., et al. (2019). Toward understanding the impact of artificial intelligence on labor. *Perspective, 116*(14), 6531–6539. https://doi.org/10.1073/pnas.1900949116

Glaser, B. G., & Strauss, A. L. (1965). Discovery of substantive theory: A basic strategy underlying qualitative research. *American Behavioral Scientist, 8*(6), 5–12.

Jarek, K., & Mazurek, G. (2019). Marketing and artificial intelligence. *Central European Business Review, 8*(2), 46–55. http://dx.doi.org/10.18267/j.cebr.213

Karjalainen, T. M., & Snelders, D. (2010). Designing visual recognition for the brand. *Journal of Product Innovation Management, 27*(1), 6–22.

Kietzman, J., Paschen, J., & Treen, E. (2018). Artificial intelligence in advertising: How marketers can leverage artificial intelligence along the consumer journey. *Journal of Advertising Research, 58*(3), 263–267. https://doi.org/10.2501/JAR-2018-035

KLM. KLM builds booking and packing bot 'BB' with Dialogflow. *Dialogflow*. Retrieved from: https://cloud.google.com/dialogflow/docs/case-studies/klm/KLM.pdf

Langfelder, N. (2020). *4 case studies on how machine learning is helping retailers drive revenue*. Retrieved from: https://www.data-axle.com/resources/blog/4-case-studies-on-how-machine-learning-is-helping-retailers-drive-revenue/

Mabert, V. A., Muth, J. F., & Schmenner, R. W. (1992). Collapsing new product development times: Six case studies. *Journal of Product Innovation Management, 9*, 200–212.

Marr, B. (2021). *What is the impact of artificial intelligence (AI) on society?* Retrieved from: https://bernardmarr.com/what-is-the-impact-of-artificial-intelligence-ai-on-society/

Mastorakis, G. (2018). *7 real world examples of how brands use artificial intelligence in marketing*. Retrieved from: https://www.mentionlytics.com/blog/7-real-world-examples-of-how-brands-use-artificial-intelligence-in-marketing/

Murgai, A. (2018). Transforming digital marketing with artificial intelligence. *International Journal of Latest Technology in Engineering, Management & Applied Science (IJLTEMAS), 7*(4). Retrieved from: https://fardapaper.ir/mohavaha/uploads/2019//Fardapaper-Transforming-Digital-Marketing-with-Artificial-Intelligence.pdf

Petropoulos, G. (2018). The impact of artificial intelligence on employment. *Bruegel*. Retrieved from: https://www.bruegel.org/wp-content/uploads/2018/07/Impact-of-AI-Petroupoulos.pdf

Pieters, W. (2010). Explanation and trust: What to tell the user in security and AI? *Ethics and Information Technology, 13*(1), 53–64. Retrieved from: https://link.springer.com/content/pdf/10.1007/s10676-010-9253-3.pdf

Rabby, F., & Hassan, R. (2021). Artificial intelligence in digital marketing influences consumer behavior: A review and theoretical foundation for future research. *Academy of Marketing Studies Journal, 25*(5), 1–7. Retrieved from https://www.proquest.com/scholarly-journals/artificial-intelligence-digital-marketing/docview/2562950050/se-2?accountid=28903

Silverman, D. (2000). *Doing qualitative research*. Sage Publications.

Wilson, H. J., Daughtery, P., & Morini-Bianzino, N. (2017). The jobs that artificial intelligence will create. *MIT Sloan Management Review, 58*(4). Retrieved from: https://www.maximo.ae/media/1306/the-jobs-that-artificial-intelligence-will-create-2-1.pdf

Yin, R. K. (1994). *Case study research: Design and methods*. Sage Publications.

Zhe Jin, G. (2018). *Artificial intelligence and consumer privacy* (National Bureau of Economic Research: Working Paper Series, 24253). Retrieved from: https://www.nber.org/papers/w24253

Digital Technology Roles for COVID-19 Crisis Management: Lessons from the Emerging Countries

Prim Patanachaisiri, Anshu Saxena Arora, and Amit Arora

Abstract The Coronavirus disease (COVID-19) has taken the world abruptly since late 2019. The nature of the newfound pandemic allows the disease to spread wide and fast at an alarming rate and has raised public concerns globally. The support of digital technologies can play an important role in assisting the situation. Technologies can aid and improve every stage of the pandemic ranging from—before the contact of the disease and till after the treatment of the disease. By utilizing AI, robotics and big data, this research focuses on the new and efficient

P. Patanachaisiri · A. S. Arora (✉) · A. Arora
University of the District of Columbia, Washington, DC, USA
e-mail: anshu.arora@udc.edu

P. Patanachaisiri
e-mail: prim.patanachaisiri@udc.edu

A. Arora
e-mail: amit.arora@udc.edu

© The Author(s), under exclusive license to Springer Nature Switzerland AG 2022
A. S. Arora et al. (eds.), *Managing Social Robotics and Socio-cultural Business Norms*, International Marketing and Management Research, https://doi.org/10.1007/978-3-031-04867-8_7

methods in handling the situation. This research serves as a guideline for policymaking in digital technology roles for handling the COVID-19 global pandemic.

Keywords COVID-19 · Digital technology · Big data · AI · Robotics · Policymaking · Emerging countries

INTRODUCTION

As of late 2019, COVID-19 has caught the world unaware. According to the World Health Organization, on January 30, 2020, COVID-19 was declared a Public Health Emergency of International Concern (PHEIC) with an official death toll of 171. By December 31, 2020, this figure stood at 1, 813, and 188. Yet preliminary estimates suggest the total number of global deaths attributable to the COVID-19 pandemic in, 2020, is at least 3 million, representing 1.2 million more deaths than officially reported (WHO, 2022). While COVID-19 is not the deadliest virus, its degree in transmissibility as well as the newness of the disease made it one of our worst enemies now. The virus can spread as easily as the flu; both direct and indirect contact, including aerosol, between people within one meter, and contact with contaminated objects or secretion substances (World Health Organization, 2022). Given the newness of the virus, the signifying symptoms along with the perfect cure have not yet been completely discovered (UCI Health, 2020). According to Harvard Health Publishing, the indicating symptoms do not need to reveal themselves for the infected individual to go on a transmissible stage. In fact, infected patients who show signs of the symptom may be less likely to be a spreader in comparison to patients whose symptoms never surface as they are more conscious and most likely have taken precautions. The estimation of the transmissible period of the disease is about 14 days starting from the initial exposure of the virus (Harvard Health Publishing, 2020).

With the above challenges at hand, the roles of digital technology in tackling the situation are significant. The interconnection of digital networks including the internet of things, artificial intelligence, and big data analytics has improved the healthcare sector's efficiency; from simultaneous data collection to analytical data processing and distribution (Ting et al., 2020). Regarding the crisis management considering

COVID-19, digital technology roles in assisting can range from handling of the situation, monitoring (Ting et al., 2020), cases analysis and projection, and vaccine development to disease prevention and supplies distribution (Vashiya et al., 2020). The precedented cases of digital technology utilization in handling the pandemic situation have been presented in several countries globally. The uses of personal records (e.g., GPS data, and credit card transactions), as well as public records (e.g., surveillance camera footages) in contract tracing of infected individuals, are carried out in South Korea (Lin & Hou, 2020); meanwhile, Taiwan has combined its national health insurance database with custom immigration data in creating simultaneous alerts through mobile messages (Wang et al., 2020).

The goal of this paper is to provide an extensive overview of digital technology for crisis management, and to illustrate how these digital technologies contribute to management of global crises (e.g., COVID-19). The research addresses the following questions:

1. How can the digital technology transform the current method of COVID-19 crisis management?
2. In which roles and stages of the process can the digital technology assist the current situation?
3. What are the policy recommendations for digital technology use for effective crisis management?

The research makes several contributions. First, there is an abundance of scientific works, particularly related to technologies' use during the COVID-19 global pandemic (Vashiya et al., 2020), however not enough research is conducted in the global crisis management and the impact of digital technologies globally. Second, the paper identified key areas regarding COVID-19 situation as well as a conceptual framework of possible digital technology roles during the crisis. Finally, global case studies of digital technology usage against the pandemic are provided to clearly illustrate the benefits and limitations for future references.

The paper is organized in the following manner. Section 7.1 provides an in-depth overview of digital technologies. Section 7.2 provides literature review on digital technologies for crisis management. Section 7.3 presents a conceptual framework of digital technologies' use during the COVID-19. Section 7.4 discusses the managerial and policy implications of digital technologies and its future progress on economic, environmental, and social environments globally.

Theoretical Background

COVID-19 Situation and Difficulties

In late 2019, a new disease associates with severe acute respiratory syndrome breakout for the first time in Wuhan, China (Zu et al., 2020). The disease is commonly compared to its, 2003, counterpart, SARS, due to several similarities including the origin, structure, transmission route, and disease progression pattern. However, while SARS was contained in 8 months with 8,098 cases and 774 deaths reported in over 26 countries, the novel COVID-19 shows no sign of stopping with 80,000 cases after 2 months of the initial case (Wilder-Smith et al., 2020). As stated earlier, according to the World Health Organization, on January 30, 2020, COVID-19 was declared a Public Health Emergency of International Concern (PHEIC) with an official death toll of 171. By December 31, 2020, this figure stood at 1, 813, and 188. Yet preliminary estimates suggest the total number of global deaths attributable to the COVID-19 pandemic in, 2020, is at least 3 million, representing 1.2 million more deaths than officially reported (WHO, 2022).

COVID-19 transmitted from peer-to-peer through secretions, droplets, both direct and indirect contact, including aerosol (World Health Organization, 2022) and has been reported to be age-dependent on mortality rate as well as more severe on people with underlying health issues (Wu & Mcgoogan, 2020). Due to the similarities in transmission route with SARS, similar countermeasures that worked in, 2003, are being enforced in the current pandemic situation which includes social distancing, self-isolation, community containment along with the hygienic measures (Wilder-Smith & Freedman, 2020; Wilder-Smith et al., 2020). However, the major different issue that heightens the difficulty in the COVID-19 is the asymptomatic cases found in the pandemic.

According to Wilder-Smith et al. (2020) and Whetton et al. (2020), most infected cases can remain asymptomatic or display only mild symptoms which leads to negligence in disease diagnosis, while the infected may very well pass on the disease. The quarantine measures, as well as the self-isolation policy after the presentation of the disease, would not be in time to prevent the transmission (Wilder-Smith et al., 2020). Those whose symptoms surface can result in a more severe case; the symptoms can include severe shortness of breath, coughing blood, pale complexion, collapse (Whetton et al., 2020), fever or chills, fatigue, and loss of taste or smell (CDC, 2020). The newness of the disease has contributed to the overall impact as more and more data are still emerging and no complete discovery on the disease has been done yet (UCI Health, 2020). In addition to the physical impact of the pandemic, the mental impact on the patients, the healthcare workers as well as the society is also evident. Zhou et al. (2020) has underlined anxiety, paranoia, and stress issues that emerge due to the pandemic and the government countermeasures. Unreliable media and misleading information can exacerbate the mental impact of the situation (Venkatesh & Edirappuli, 2020).

Digital Technology Roles in the COVID-19 Situation

As mentioned above, the time-sensitivity and the communicability factors are vital in controlling the situation. The high degree of peer-to-peer transmissibility of the disease and the complexity of the situation has made the utilization of digital technology even more prominent. There are precedented cases of successful digital technology integration in different areas, for instance, self-quarantine, surveillance contact tracing (Whitelaw et al., 2020), travelers' contamination risk assessment, resource allocation (Wang et al., 2020), etc. Several works of literature identify the possible area and dimension for digital integration in the COVID-19 situation.

According to Ting et al. (2020), the already established interconnected network of digital technology has allowed simultaneous data collection, analytical data processing, and data distribution to be utilized along with two traditional strategies in handling COVID-19, which are (a)monitoring, detection, and prevention and (b) mitigation of aftermath. Accessibility to the global simultaneous data not only enhances acceleration in data sharing but also ensure the relevancy of the data that can, then, be used for modeling the trajectory as well as the course of the pandemic. Big data and AI can aid the early detection of the symptoms especially in the mild case that is easily negligible at first glance. The contactless policy can be aided by digital tools, for example, telehealth and contactless drug delivery (Ting et al., 2020). Vaishya et al. (2020) have proposed seven significant areas that can benefit from digital tools including (a) early detection, (b) treatment tracking, (c) contact tracing,

(d) cases trajectory, (e) remedies research and development, (f) healthcare workload alleviation, and (g) disease prevention (Vaishya et al., 2020).

Bragazzi et al. (2020) classify the main areas of integration in a time-based manner: short-term, medium-term, and long-term. In the short-term digital technology can respond rapidly to new cases together with generating alert and contact tracing. In the medium-term digital technology roles would be more prevalent in medical care for existing cases, research and development for possible cures, and resource allocation along with smoothing the public health intervention processes. Lastly, in the long-term, the utilization of digital technology would be focusing on the construction of the smart infrastructure to properly manage future challenges (Brigazzi et al., 2020). Other suggestions have been made in specific areas, for instance, the identification of biomarker related to the current pandemic (Whetton et al., 2020); and the implementation of technology tools to aid the mental health crisis stems from the pandemic and government policy (Zhou et al., 2020).

Conceptual Framework

The conceptual framework (Fig. 7.1) depicts the possible applications of digital technology application in the COVID-19 situation throughout the journey of a COVID-19 patient.

While some areas like data collection along with research and development of remedies occur in every stage throughout the patient's COVID-19 journey to ensure the continuous improvement of the situation and the fastest cure possible. Some other areas are more related to stages of the journey than others. The framework arranges different scenarios based on the ongoing stages of the disease.

Before: The Pre-Covid-19 Scenario

In the pre-COVID-19 stages, accessibility of accurate information is indispensable as misinformation and fake news along with unclear guideline communication led to anxiety and panic (for example, the case of panic buying of commodities), which exacerbate the situation. Clear guidelines and easy-to-access, reliable information sources that can reach all levels of populations promptly can debunk and clear up unnecessary complications in the early phase of disease control (Fagherazzi et al., 2020). Communications through authorized online channels as well as the utilization of

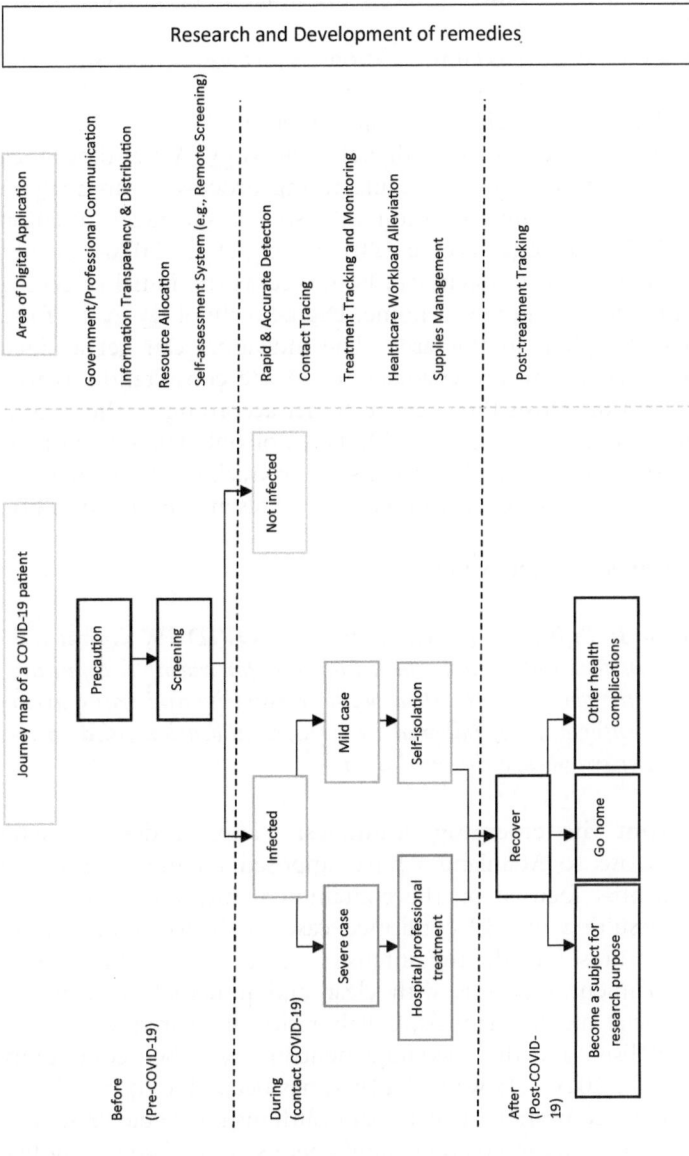

Fig. 7.1 Conceptual framework depicted possible areas of digital technology application in the COVID-19 situation throughout the journey of a COVID-19 patient

application and instant messaging services can support the quantity and immediate requirements needed to tackle the situation (Fagherazzi et al., 2020; WHO, 2020).

Following the guidelines communications, shortage of necessities and resources that are essential for preventive measures against COVID-19 (for example, face masks, gel hand sanitizer, etc.) (Southey, 2020) are the next issue that can be resolve by digital technology. A real-time stock tracking system, as well as smart manufacturing processes, can analyze, identify, and make a prompt decision in resource shortages situation (Javaid et al., 2020; Manogaran et al., 2017; Zeng et al., 2020). Finally, the self-assessment support system can be served as the initial screening stage that can be done remotely at home. Digital technology in the form of telehealth is prevalent in this area. The individual can get a video conference with healthcare professionals to do the consultation as well as initial symptom diagnosis they needed to act according to the healthcare guidelines (Fagherazzi et al., 2020). This not only takes care of the physical dimension of the situation but also keep track of the mental side as an individual now have access to reliable services at their hand (Zhou et al., 2020).

Thus, we propose the following:

> **Proposition I:** *Before the patient contracts COVID-19, digital technology role in terms of reliable media outlets (for example, communication from the government) along with accessibility and transparency in essential information, followed by resource allocation, and initial self-assessment system are the vital areas.*

Example from the emerging countries: Vietnam, despite being geographically close to Mainland China, imposed countermeasures in the initial stage that focused on the containment measures against the novel disease, resulting in 239 confirmed cases with no loss reported. The countermeasures include the intensive screening of information, the rapid government responds with clear and punctual communication through social media channels, collaboration in shared knowledge in science journalism, and the encouragement in strengthen community network (La et al., 2020). Taiwan, similarly, has been able to handle the situation rapidly by combining its national health insurance database with custom immigration data in creating simultaneous alerts through mobile messages. Government communications were made clear and accessible through the internet. Resource management, as well as logistic control, are supported by the government to ensure fair distribution (Wang et al., 2020).

DURING: THE AFTER INFECTED SCENARIO

The during-phase of COVID-19 is the stage in which individuals who are diagnosed with the infection will need to experience. To transit smoothly through this stage, the utilization of digital technology in accurate disease detection along with contact tracing of individuals is especially eminent. As the disease is very time-sensitive due to its degree in transmissibility, the earlier the detection, the more efficient in the containment of the disease. AI and big data can aid in rapid and accurate diagnosis (Whetton et al., 2020) while rapid contact tracing can quickly identify risk groups, generating alerts, and allow for on-point government intervention, including quarantine order (Hellewell et al., 2020). Personal and public digital records like GPS data, credit card transaction records, data from surveillance cameras can be contributed to the effectiveness in contact tracing (Lin & Hou, 2020). The treatment tracking and monitoring areas are vital in both mild and severe cases of infection. While technology can aid the self-isolation process in mild cases, the use of AI and robotic in hospitals or professional facilities not only help with data collection and visualization but also healthcare workload and risk alleviation (Javaid et al., 2020). All of which will contribute to the more efficient pandemic containment and treatment. Thus,

Proposition II: *After COVID-19 is contracted, digital technology role in terms of early detection, contact tracing of individuals, treatment tracking and monitoring, healthcare workload alleviation, and supply management are the vital areas.*

Example from the emerging countries: South Korea and Taiwan have done an exceptionally well job in utilizing digital technology in this stage. The early detection along with contact tracing of individuals was succeeded by proactively seeking out cases throughout all possible hotspot identified by the help of personal and public records, including surveillance footages, credit card transaction records, GPS data (Lin & Hou,

2020), information from border control, and travel history (Wang et al., 2020). The monitoring of self-isolation and quarantine which helps alleviate the workload of healthcare sectors are also made possible through mobile data tracking (Lin & Hou, 2020).

After: The Post-Covid-19 Scenario

In the post-COVID-19 stages, the patient has recovered from the disease, and in some cases, their plasma is used for further research purposes (CDC, 2020); while in some other cases, the disease has left other health complications issues (Fraser, 2020). The aftercare of recovered patients as well as the tracking of possible health complications, including psychological impact, are important areas of focus to ensure a healthy and complete recovery. Personal tracking applications and telehealth can aid in this area. Thus,

> **Proposition III:** *After the patients recover from COVID-19, digital technology can emphasize on the post-recovery tracking of individuals.*

Discussions and Conclusions

With such a high degree of challenges at hand, countries' leaders and policymakers need to mitigate the situation as well as minimizing loss as much and as fast as possible. Propositions I—III provide cases and scenarios regarding how digital technology can be utilized before, during, and after COVID-19 contraction leading to resource allocation, initial self-assessment system, early detection system, contact tracing of individuals, treatment tracking and monitoring, healthcare workload alleviation, supply management of medicines and healthcare, and post-recovery systems/tracking of patients recovering from COVID-19 pandemic.

While the need to find the remedies is eminent, maintaining all other aspects of the situation is no less important issue. The theoretical framework helps illustrate, identify, and classify each significant area of focuses regarding different scenarios in the COVID-19 journey of individuals. There are different focal points in each stage of the journey where the utilization of digital technology can be a part of (for example, telehealth for initial screening and self-assessment). The example from emerging countries can serve as case studies in the implementation of digital technology as assistant tools in tackling the situation at hand. The integration and viability of digital technology in each region can be different as there are various degrees of information security concerns; nonetheless, digital technology roles will still be there, oncoming to prepare for future challenges.

References

Adams, V. J., Azar, A., Fauci, A., Giroir, A. B., Hahn, S., Redfield, R., & Woodcock, J. (2020). *Donate blood plasma and help save lives.*

Anderson, R. M., Heesterbeek, H., Klinkenberg, D., & Hollingsworth, T. D. (2020). How will country-based mitigation measures influence the course of the COVID-19 epidemic? *The Lancet, 395*(10228), 931–934.

Bragazzi, N., Dai, H., Damiani, G., Behzadifar, M., Martini, M., & Wu, J. (2020, May 2). *How big data and artificial intelligence can help better manage the COVID-19 pandemic.* Retrieved September 13, 2020, from https://www.mdpi.com/1660-4601/17/9/3176/htm

Fagherazzi, G., Goetzinger, C., Rashid, M. A., Aguayo, G. A., & Huiart, L. (2020). Digital health strategies to fight COVID-19 worldwide: Challenges, recommendations, and a call for papers. *Journal of Medical Internet Research, 22*(6), e19284.

Fraser, E. (2020). *Long term respiratory complications of covid-19.*

Hellewell, J., Abbott, S., Gimma, A., Bosse, N. I., Jarvis, C. I., Russell, T. W., ... & Flasche, S. (2020). Feasibility of controlling COVID-19 outbreaks by isolation of cases and contacts. *The Lancet Global Health.*

Jason Wang, C. (2020, April 14). *Response to COVID-19 in Taiwan: Big data analytics, new technology, and proactive testing.* Retrieved September 13, 2020, from https://jamanetwork.com/journals/jama/article-abstract/2762689

Javaid, M., Haleem, A., Vaishya, R., Bahl, S., Suman, R., & Vaish, A. (2020). Industry 4.0 technologies and their applications in fighting COVID-19 pandemic. *Diabetes & Metabolic Syndrome: Clinical Research & Reviews.*

La, V. P., Pham, T. H., Ho, M. T., Nguyen, M. H., Nguyen, K. L., Vuong, T. T., ... & Vuong, Q. H. (2020). Policy response, social media and science journalism for the sustainability of the public health system amid the COVID-19 outbreak: The Vietnam lessons. *Sustainability, 12*(7), 2931.

Lin, L., & Hou, Z. (2020). Combat COVID-19 with artificial intelligence and big data. *Journal of Travel Medicine, 27*(5), taaa080.

Manogaran, G., Thota, C., Lopez, D., & Sundarasekar, R. (2017). Big data security intelligence for healthcare industry 4.0. In *Cybersecurity for Industry 4.0* (pp. 103–126). Springer.

Publishing, H. (2020, March). *If you've been exposed to the coronavirus.* Retrieved September 24, 2020, from https://www.health.harvard.edu/diseases-and-con ditions/if-youve-been-exposed-to-the-coronavirus

Q&A: How is COVID-19 transmitted? (n.d.). Retrieved September 24, 2020, from https://www.who.int/emergencies/diseases/novel-coronavirus-2019/ question-and-answers-hub/q-a-detail/q-a-how-is-covid-19-transmitted?gclid= CjwKCAjwh7H7BRBBEiwAPXjadn5K7WQQgY5f3dk7__IQBOxxY4dSSd IeTTr8hgdBoEfQzqHGaI7bwxoCDIgQAvD_BwE

Similarities and Differences Between Flu and COVID-19. (2020, August 31). Retrieved September 24, 2020, from https://www.cdc.gov/flu/symptoms/ flu-vs-covid19.htm

Southey, F. (2020). Panic buying amid coronavirus fears: How much are we spending... and why is it a problem. *Food Navigator.*

Ting, D., Carin, L., Dzau, V., & Wong, T. (2020, March 27). *Digital technology and COVID-19.* Retrieved September 13, 2020, from https://www.nature. com/articles/s41591-020-0824-5

UCI Health. (2020, April 29). *Why is COVID-19 do dangerous?* Retrieved September 21, 2020, from https://www.ucihealth.org/blog/2020/04/why- is-covid19-so-dangerous

Vaishya, R., Javaid, M., Khan, I. H., & Haleem, A. (2020). Artificial Intelligence (AI) applications for COVID-19 pandemic. *Diabetes & Metabolic Syndrome: Clinical Research & Reviews.*

Venkatesh, A., & Edirappuli, S. (2020). Social distancing in covid-19: What are the mental health implications? *Bmj, 369.*

Wang, P. W., Lu, W. H., Ko, N. Y., Chen, Y. L., Li, D. J., Chang, Y. P., & Yen, C. F. (2020). COVID-19-related information sources and the relationship with confidence in people coping with COVID-19: Facebook survey study in Taiwan. *Journal of medical Internet research, 22*(6), e20021.

Whetton, A. D., Preston, G. W., Abubeker, S., & Geifman, N. (2020). Proteomics and informatics for understanding phases and identifying biomarkers in COVID-19 disease. *Journal of Proteome Research.*

Whitelaw, S., Mamas, M., Topol, E., & Spall, H. (2020, June 29). *Applications of digital technology in COVID-19 pandemic planning and response.* Retrieved September 13, 2020, from https://www.sciencedirect.com/science/article/ pii/S2589750020301424

World Health Organization, WHO. (2022). *The true death toll of COVID-19: Estimating global excess mortality.* Retrieved January 26, 2022. https://www.who.int/data/stories/the-true-death-toll-of-covid-19-est imating-global-excess-mortality.

Wilder-Smith, A., Chiew, C. J., & Lee, V. J. (2020). Can we contain the COVID- 19 outbreak with the same measures as for SARS? *The Lancet Infectious Diseases.*

Wilder-Smith, A., & Freedman, D. O. (2020). Isolation, quarantine, social distancing and community containment: pivotal role for old-style public health measures in the novel coronavirus (2019-nCoV) outbreak. *Journal of Travel Medicine, 27*(2), taaa020.

Wu, Z., & McGoogan, J. M. (2020). Characteristics of and important lessons from the coronavirus disease 2019 (COVID-19) outbreak in China: Summary of a report of 72 314 cases from the Chinese Center for Disease Control and Prevention. *JAMA, 323*(13), 1239–1242.

Zeng, J., Huang, J., & Pan, L. (2020). How to balance acute myocardial infarction and COVID-19: The protocols from Sichuan Provincial People's Hospital. *Intensive Care Medicine, 46*(6), 1111–1113.

Zhou, X., Snoswell, C. L., Harding, L. E., Bambling, M., Edirippulige, S., Bai, X., & Smith, A. C. (2020). The role of telehealth in reducing the mental health burden from COVID-19. *Telemedicine and e-Health, 26*(4), 377–379.

Zu, Z. Y., Jiang, M. D., Xu, P. P., Chen, W., Ni, Q. Q., Lu, G. M., & Zhang, L. J. (2020). Coronavirus disease 2019 (COVID-19): A perspective from China. *Radiology*, 200490.

CHAPTER 8

Social Commerce: Impact on Consumer Power Through Social Media

*Jaehwan Kim, Curley Ohio, Prim Patanachaisiri,
Renata Buzolin, Carla Jamille Arapiraca,
and Anshu Saxena Arora*

Abstract The increasing penetration of commerce through new technologies is changing consumer behavior from analog to digital. Social
media facilitate consumers' real-time interactions allowing them to access
a vast amount of information thereby increasing their consumer power in
the virtual environments. On the other hand, many brands utilize "social

J. Kim · C. Ohio · P. Patanachaisiri · R. Buzolin · C. J. Arapiraca ·
A. S. Arora (✉)
University of the District of Columbia, Washington, DC, USA
e-mail: anshu.arora@udc.edu

J. Kim
e-mail: jaehwan.kim@udc.edu

C. Ohio
e-mail: curley.ohio@udc.edu

P. Patanachaisiri
e-mail: prim.patanachaisiri@udc.edu

© The Author(s), under exclusive license to Springer Nature
Switzerland AG 2022
A. S. Arora et al. (eds.), *Managing Social Robotics and Socio-cultural
Business Norms*, International Marketing and Management Research,
https://doi.org/10.1007/978-3-031-04867-8_8

commerce" to promote and sell their products but are still struggling in effectively using computer-mediated social environments (CMSEs) in order to create a network approach to branding. Therefore, this chapter aims to better understand consumer behavior and consumer power in physical and virtual worlds, as well as, comprehend how to effectively engage in social commerce for improving brand management and sales by successfully connecting consumers. Based on the three case studies of Uber, Pinterest, and Airbnb, we analyze three propositions.

Keywords Consumer power · CMSEs · Brand management · Social media · Consumer · Decision-making process · Consumer behavior · Physical versus virtual environments

INTRODUCTION

Nowadays, the strategies that marketers use to enhance social commerce in computer-mediated social environments (CMSEs) must be reinvented several times to progress with the technology advances. The technical infrastructure has transformed over the past decade, especially the development of information technology (IT), which makes people's lives more convenient. IT can have both positive and negative effects on individuals' behavior, organizations, environments, and markets, especially on CMSEs, which is the networking of websites such as Facebook, Instagram, Twitter, Pinterest, and others, with the purpose of connecting people. The term social commerce is used to refer to people who use social media for commerce purposes, like search for information about a certain product, or even buy and sell goods.

As mobile technology is now considered the most aggressive communications medium (Nafees et al., 2021; Ng & Vranica, 2013), marketers

R. Buzolin
e-mail: renata.buzolin@udc.edu

C. J. Arapiraca
e-mail: carlajamille.arapira@udc.edu

spend more money on it than on TV, and similarly, consumers spend more hours on their cell phones (eMarketer, 2013). Although consumers have the ease of accessing information, they also have the issue of the distraction (Birkner, 2015; Griffith et al., 2022). Branding has become a vital asset for firms and managers (Keller, 1993), especially with the emergence of social media; consumers are now empowered to share their brand stories widely via social networks (in real-time) which has impacted on consumer purchasing behavior/experiences (Kuksov et al., 2013).

Some consumers integrate both virtual and physical environment when comparing products, purchasing something and sharing their experiences and/or level of satisfaction with other consumers. Knowing that, we believe it is important to understand the consumer behavior in each stage of the decision-making process in both environments. This consumer behavior links to consumer power, which is raised in CMSEs. Companies build a good relationship with consumers by answering inquiries, sending promotions, and sharing new product and service information. However, consumers are sometimes relying on negative reviews. For example, an influencer makes a bad review, consumers would not purchase the product based on the review.

Consumer power in the online environment has been rising with the advance of technologies. People have various expressions of consumer behavior in digital media such as influencing others, sharing information, and impacting on decision-making (Labrecque et al., 2013). More research is needed about why social media attracts consumers even though academics have been trying to understand such events. Also, it is significant to examine how marketers could increase goods and services sales making use of the computer-mediated social environments, which includes social media (Griffith et al., 2022; Hoffman et al., 2013; Yadav et al., 2013).

There is a dearth of research in the field of social commerce and consumer power in CMSEs (Banerjee & Longstreet, 2016; Gensler et al., 2013; Griffith et al., 2022; Labrecque et al., 2013; Nafeez et al., 2021; Yadav et al., 2013). To bridge the gap, our research has the following objectives.

- Examine the differences (and similarities) in consumer behavior prevalent in dual physical-virtual environments through the consumer decision-making process;

- Analyze how social media affects consumer power and consumer interactions in CMSEs;
- Evaluate how social commerce can help in brand management; and
- Investigate how marketers can effectively use social commerce to increase sales and branding.

Our research consists of four sections. First, we focus on defining and describing the differences among the types of consumer behavior and consumer power through social commerce. Second, we examine how social media has impacted, and continues to impact, the field of marketing and the effects it has on consumer and business behavior. Next, we utilize the Consumer Power-based Social Commerce (CPSC) framework to explore the benefits and risks associated with utilizing the consumer power in dual CMSEs. Managers will be able to assess their current advertising platforms by taking advantage of our framework, deciding whether to increase social media advertising or remain consistent with the static approach. Additionally, we discuss the positive influences that CPSC (through utilization of social media in CMSEs) framework will have on business-to-business relationships, as well as business-to-consumer relationships.

THEORETICAL BACKGROUND

Dual Environments and CMSEs

Dual Environments, according to Banerjee and Longstreet (2016), represent any given physical space used by someone while this person is having access to a mobile device connected to the internet as well. This interaction of an individual with both environments has two main outcomes, which are integration or dissociation.

Integration happens when a person has a high interaction with both physical and virtual environments, complementing one another. For example, integration can happen when a consumer compares prices online with prices in physical stores. This way, they can improve their cost-effective shopping and make the right decision (Banerjee & Longstreet, 2016; Griffith et al., 2022).

Dissociation, different than integration, is when a person has a low interaction with one of the environments, virtual or physical, while having a high interaction with the other one. For instance, someone who drives and texts at the same time has a high interaction with the virtual environment and a low interaction with the physical one. Most of the cases of dissociation lead to bad endings, such as injury, discord, and lawsuit or, in extreme cases, death (Banerjee & Longstreet, 2016).

Computer-mediated social environments (CMSEs) are considered the integration of a network with associated hardware and software with social media that allows people to interact and share and search for information with each other. We can name famous social media like Facebook, Twitter, and Pinterest that have emerged to the computer-mediated environment (CME) with social characteristics, making the CME social (Yadav et al., 2013).

As mentioned above, integration is a successful result of a dual environment experience, while dissociation results in something negative. Consumers tend to make use of dual environments when buying and/or sharing experiences about a product, or even when looking for helpful information about that product like prices, options and recommendations from other consumers. It is easy and convenient for consumers to have a vast number of possibilities—included in both physical and virtual environment—when going through the decision-making process.

Several brands make use of social media—virtual environment—to advertise and sell their products, and that is what we call social commerce: commercial activities that happen online or are influenced by an online environment (Griffith et al., 2022; Yadav et al., 2013). For instance, a consumer can see an advertisement on his Twitter account and proceed with the purchase exclusively online, which can include activities such as searching for information regarding the product, like prices and other matters, and finally the actual purchase. However, this consumer can totally choose making use of dual environments in this case by looking at that virtual advertisement on Twitter and realizing the purchase in a physical store, resulting in integration—a positive result from the use of dual environments.

Given the above discussion, we can say that social commerce is a function of social media, which means that it just happens because people use social media not only to connect with each other but also to realize commercial activities—what we call social commerce—and this fact impacts dual CMSEs in a way that consumers can lead to integration or dissociation when making use of dual environments when purchasing goods or services.

Consumer Behavior and Social Commerce in Dual CMSEs

Consumer Power, according to Labrecque et al. (2013), is the intersection of consumer behavior and digital media by clearly defining consumer power and empowerment in Internet and social media contexts and by presenting four consumer power frameworks: demand, information, network, and crowd-based power. Therefore, consumers can obtain and share any kind of information in the virtual environment, influencing purchase decisions, impacting website visits, and increasing application downloads.

Social Commerce is any exchange-related activity that happens in a virtual environment or, at least, is influenced by social media. In other words, it is when an individual go through the decision-making process using the virtual environment or combining it with the physical environment (Griffith et al., 2022; Yadav et al., 2013). We can conclude that it is important to think about both physical and virtual environments when creating marketing strategies, since a purchase can be influenced by a certain social media as well as by the physical environment.

The research examines how consumer power impacts and interacts with social commerce. The number of social connections in one's network substantially increases consumer power, which is the ability to share and influence others. Consumers tend to co-create content more easily through liking, commenting, tagging, or other forms of social media function. For example, consumers share their contents using the social media functions, and then the contents will spread wildly in the social networking sites. A study explains how information spreads through social and online activities, identifying four distinct communication strategies: evaluation, embracing, explanation, and endorsement (Labrecque et al., 2013). These strategies are used by social media influencers to spread product information and, consequently, the readers become another influential consumer in a network. This network-based consumer power impacts on social commerce market due to exchanging information.

Network Approach to Branding

This approach insinuates that a consumer relationship with a brand occurs through consumer's social connections, leading to impact on other consumers by such social connections related to the brand. Hence, brand and customer relationship management are fundamentally connected in a social media environment (Malthouse et al., 2013). Network-oriented approach to branding makes it clear that a consumer purchasing power of a product is affected by his/her interactions on social media platforms. This happens because consumers bring value to products/brands based on the reviews and comments that they make on their social media, determining future consumer decisions about whether to purchase that determined product. Alternatively, a consumer who does not purchase too much from a specific brand might still influence others in the virtual environment (Griffith et al., 2022; Kumar et al., 2013). Hence, "a brand is no longer what we tell the consumer it is - rather it is what consumers tell each other it is" (Gensler et al., 2013, p. 242).

Furthermore, the use of social media has dramatically changed the way firms manage their brands. As a brand's social network now consists of many voluntary connections of consumers (e.g., people who voluntarily follow or like a brand online page), this affects the authenticity of a brand's social identity and, at the same time, adds complexity to the management of brand identity (Naylor et al., 2012). Social media, in essence, has remarkably increased the reach and visibility of consumer social networks and makes it effortless to mobilize consumers (e.g., on Twitter, Facebook, and others), causing positive impacts on social commerce (Kane, 2009).

CONCEPTUAL FRAMEWORK

Consumer Power-based Social Commerce (CPSC) Framework

As we progress toward an era of exponential technology growth and development, people propensity change from physical to virtual environments. They can easily access the Internet, exchange information, discover and adapt new ideas, and influence others, which in turn will positively impact market share, marketing, and/or business strategies.

This research explores the relationship between consumer and dual environments to explain the Consumer Power-based Social Commerce (CPSC). At first, we realize a study about how consumers interact with both physical and virtual environments at the same time. After that, we investigate who is the consumer and what are these environments. Next, we analyze the two outcomes resulted from that interaction, defined by integration and dissociation. Then, the research leads us to understand the consumer power and its three frameworks: power, influence and empowerment. This power is exemplified by the network crowd power, going through its creation, funding, sourcing, selling and support. Figure 8.1 illustrates our CPSC framework in CMSEs.

The next phase of this research is narrowed to computer-mediated social environments (CMSEs), giving an understanding of what social commerce is, following the stages of consumer decision-making process. With all this information explored here, we can find solutions for brand

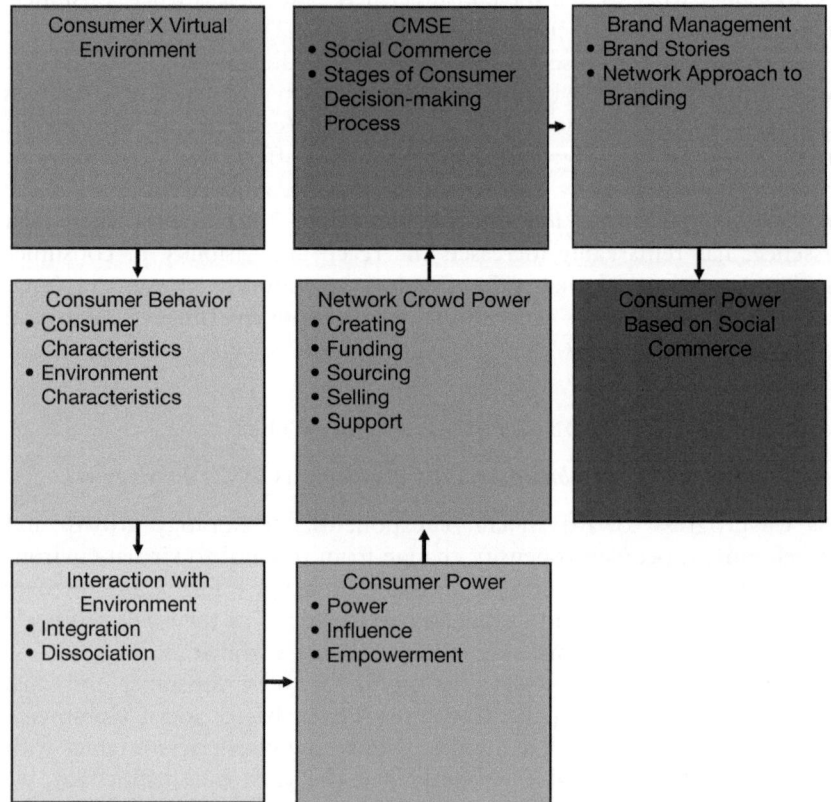

Fig. 8.1 CPSC Framework in CMSEs: How "Consumer Power" and "Social Commerce" Impact "Brand Management," and "Consumer Power-based Social Commerce (CPSC)"

management and sales improvement, using brand stories and network approach to branding. Finally, we find our new concept defined as Consumer Power-based Social Commerce (CPSC).

Dual Environments

Dual environments are known as distracters of consumer's mind, which is taken because a lot of people dissociate when in situations that include physical and virtual environments. For retailers and marketers, this could be a problem when it is taking the consumer's attention away from their product (Banerjee & Longstreet, 2016). On the other hand, some consumers are seeking for an improved experience and integrate the environments using tools like QR, to retrieve information about the product displayed, however, some consumers could make the use of Price Compared Decision-Making (Banerjee & Longstreet, 2016), impairing the retailers. So, the integration or dissociation could both affect the sales process, and the Consumer Power-based Social Commerce is a matter of understanding who the consumer is and how to reach them to solve this issue. In conclusion, CPSC in dual environments affect the decision-making process because it helps consumers to find the best option of purchase, not making them loyal to any specific brand or provider. Thus, we offer the following proposition.

Proposition I: Dual environments affect consumer purchase and decision-making process.

Consumer Power

Consumers tend to look for product information online, from other consumer reviews before making a decision. Such fact relates to consumer power in the virtual environment, and in the physical environment if they decide to continue the decision-making process out of the virtual network. As an example, we can use the Airbnb business, which is an online marketplace that allows people to rent their spare rooms to other people by sharing information in the virtual environment. Due to high demand of positive consumer experiences and new IT technology, Airbnb has become a company valued at $2.5 billion in six years (Kessler, 2014). Besides that, consumers can make online reviews so other consumers that are looking for information about that specific host are able to better analyze the options based on experiences posted online. Because of that, the Airbnb Company has improved customer reputation system based on negative reviews by adding features to the review system (Edelman & Luca, 2012). This exemplifies how consumer power positively impacts social commerce since consumers can exchange experiences and information, helping each other in dual CMSEs. To conclude, the impact that consumer power has over social commerce, what we define by consumer power-based social commerce (CPSC), can easily happen in dual environments and cause positive results. Thus, we offer the following proposition.

Proposition II: When consumer power increases, social commerce increases as well in CMSEs.

Brand Management

Social commerce increases as consumer power increases, forming a dual relationship, since they are directly related. In addition, it is relevant to remember that many brands make use of dual environments to increase sales. A good case study to support this proposition is the company Yelp, which is a website and third-party platform devoted to providing reviews about services like restaurants, hotels, gas stations, and others (Yin et al., 2014). Because of that, consumers tend to use this platform to search for information before making the actual purchase, and Yelp has become a very useful tool in serving this purpose. A good review is perceived to be helpful to influence others purchase power. On the other hand, a negative review can affect consumer purchasing decisions (Yang & Mai, 2010). That is the reason marketers should pay close attention to reviews in dual CMSEs, to improve brand management by benefiting from consumer power since it impacts social commerce and businesses profits. Thus, we offer the following proposition.

Proposition III: As social commerce activities increase, consumer power increases as well as making business profit through better brand management in dual CMSEs.

CASE STUDY METHODOLOGY

The objective of this study was to examine consumer power, social commerce, and brand management through CPSC conceptual framework, and how CPSC affects both consumers and businesses. To conduct the research, several case studies were engaged, which is generally considered more robust than single case study (Yin, 1994). The cases were deliberately selected, and the case methodology presented here is consistent with the objectives of qualitative research (Glaser & Strauss, 1965; Silverman, 2000). The research methodology follows closely to qualitative works including Brockman et al. (2010) and Mabert et al. (1992) that utilize case research to drive new framework or theory. Our qualitative research includes CPSC framework related variables (as explained above), and we investigate these variables through an in-depth analysis of three case studies.

Uber and the Sharing Economy

The type of business that Uber follows can be included in the sharing economy model, which is used for travel, car sharing, finances, staffing, music, and video. Uber is a crowd-based consumer power due to its advances in mobile technology and data infrastructures which allow for the rise of group/community buying power and the sharing economy. Social media created access to information for consumers and the Internet has been making the sharing economy expand, although it's an old concept. This economic model is positive in terms of sustainability and social benefit, once it is led by the sense that is important to promote interactions, meaningful connections, and reduce waste by making maximum use of goods' capacity.

The Uber Company provides a technology platform that allows customers to receive easy and trustful access to a ride using an app on their mobile devices. It is considered a lifestyle company that creates value for their customers by giving them the opportunity to experience things like having a car with a simply push on a button. The company has specialized services like UberHealth, UberAssist, UberRush, and UberEats. Pricing is an important component of this business, and their primary appeal is that drivers could earn extra money as an independent professional using their own vehicle, make their own schedule, and receive weekly payments. We can say that Uber customers make use of dual environment when

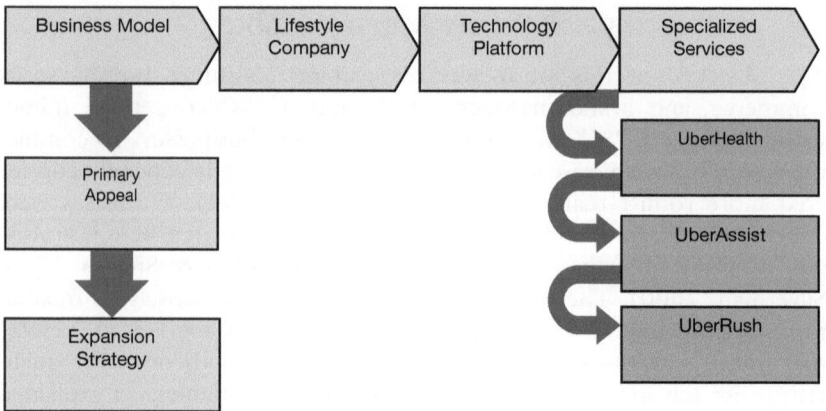

Fig. 8.2   The Uber Company

buying the service, since they order and pay the rides online, however, the actual service happens in the physical environment, leading to integration. Figure 8.2 illustrates the Uber business.

The business is doing well in the United States; however, it's facing big competitors in Europe and Asia. Knowing that, the company has an expansion strategy, which includes raising money for expansion of projects and implementing the sharing economy model city-by-city launches in target areas. Figure 8.3 shows that in the United States Uber competes with Lyft, that offers basically the same kind of service, however does not have international presence; Zipcar, which is a different business model where consumers are able to reserve a car online, pick it up and drop it off at designated locations, by paying an annual subscription, and this company operates internationally as well; and finally they

Domestic	International
Lyft	Europe - Taxi
Zipcar	China - Didi Kuaidi
Taxi	India - Olacab

Fig. 8.3   Uber competitors

compete with the Taxi industry, which has a rigorous licensing process making it sounds safer than Uber, leading the market share for that reason. In Europe, China, and India, Uber competes with the respective companies/industries: taxi, Didi Kuaidi and Olacab. The company is having a difficult time to gain consumers trust and increase profits. For instance, Olacab sold 750,000 rides on a certain day, while Uber sold only 280,000.

Pinterest

Pinterest, Inc. was co-founded by Ben Silbermann, Paul Sciarra, and Evan Sharp in December 2009. It is an American-based social media platform and mobile application company, with its headquarters located in San Francisco, United States. It is designed essentially to enable people to gather, save, organize, and share images, videos, graphic interchange formats (GIFs), and other information relevant to specific areas of interest on the World Wide Web (www) by the use of the Pinterest bookmarklet. Although the website is free to access, it requires signing. As of August 2019, Pinterest active users have hit 300 million (Fiegerman, 2019). By way of definition, Pinterest is a social media platform, as well as a strong search engine/function. It enables users to visually save, share, and uncover latest interests by pinning videos or images to their board (this is a collection of "pins") or other peoples' boards (Meng, 2019). Pinterest operates both personal and business accounts. But you can easily convert your personal account into a business account. Moreover, Pinterest permit businesses to create pages focused at promoting their organizations online. In fact, you can find almost anything on Pinterest, ranging from Art to Technology, even to Dinner Recipes. Pinterest witnessed a dramatic growth in 2013; transcend email as a sharing forum, and even exceeded Facebook. According to ShareThis (an online content dissemination resource), Pinterest is "now the fastest-growing forum for online content distribution."

Some Pinterest Terminologies

Pins—A pin is simply an image connected from a website, so that when you click on it, it automatically links you to the website.

Pincodes—They are unique codes you generate to open your business categorized boards and profile; they operate just like QR codes.

Boards—These are basically collections of pins tailored to specific themes or ideas; they function like digital bulletin boards.

Pinners—These are essentially the people who make use of Pinterest.

Feed—This resembles a Twitter or Facebook feed, consisting of pins from people you follow or that Pinterest presume you may be interested in.

Lens—This distinctive attribute on Pinterest makes use of your phone's camera to create pin suggestions in line with the items you photographed.

Rich Pins

They are remarkable pins that make operating the platform effortless and coherent. These unique pins contain information that may not be displayed on the image, such as the pricing information which makes them quintessential if you are operating a business account. There are 4 main types of Rich Pins, namely: App Pins, Product Pins, Article Pins, and Recipe Pins (Saige-Driver, 2018).

The phenomenon of showrooming, where people visit the stores to check on items and end up buying them online, is worrying brick and mortar retailers as they believe sales are being affected by that. This case study is analyzing how Pinterest affect sales and if that belief is accurate. When the results of the research made were available, it was observed that showrooming did not affect sales as retailers thought it was, only 26% of interviewed people declared doing it. On the other hand, 41% of the interviewed people admitted practicing the opposite of showrooming, what is called reverse showrooming. The users of Pinterest usually find the items online and then go to the store to check on it in person and after that they buy it. Therefore, Pinterest is a network and crowd-based consumer power because it utilizes social networks and the development of mobile technology to retrieve information and effectively distribute its contents. There are five types of seekers identified among the Pinterest users as shown in Fig. 8.4.

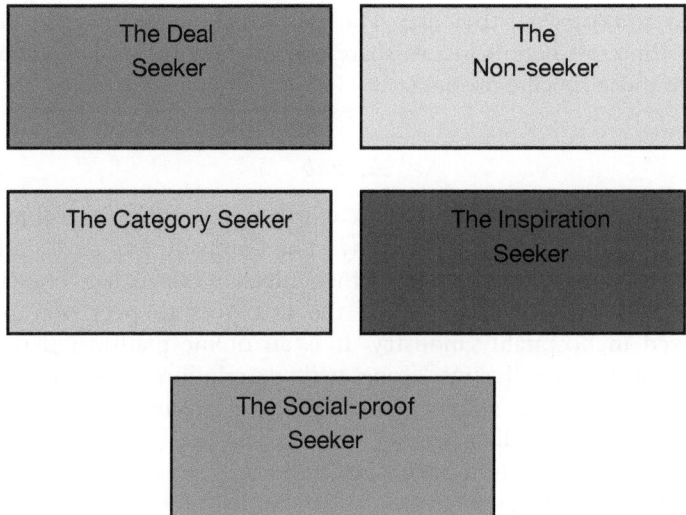

Fig. 8.4 The five types of seekers

- **The Deal Seeker** is the user that is always trying to find the best deals, and it is seeking to shop. These users are influenced by what they see on Pinterest, and they are always paying attention to other websites, emails, and others;
- **The Non-seeker** is the type of user that is not focused on trying to find anything specific, instead they are just exploring the website as they go. Anything can lead to a potential effective shopping for them in case they like something they see. These users are influenced by what they pin;
- **The Category Seeker** is someone that uses the Pinterest to gather pins about a specific subject they like, i.e., music. The role of Pinterest here is to inform about the new things in this category and the pins could remind them to buy something they want;
- **The Inspiration Seeker** is looking for ideas from other consumers that could inspire them to purchase a certain product. These users are also not trying to find anything specific; however, they are affected by what they pin and influenced on their shopping power;
- **The Social-Proof Seeker** differs from the others because they go on the opposite direction; they usually already know what they want to

buy and/or what they like. The reason behind pinning the pictures on Pinterest here is just to share and to show their friends that they like those specific products.

Airbnb

Airbnb is another company that is worth mentioning regarding social commerce and the sharing economy. The company was co-founded by Brian Chesky, Joe Gebbia, and Nathan Blecharczyk in San Francisco in 2008, where they became one of the first peer to peer services that specialized in hospitality industry. It is an online platform that allows hosts to post room listings, along with descriptions and photographs, and the guest could search for rooms in a given area and examine possible options including reviews, descriptions, and photographs. Before accepting a reservation, a host also could see the profile of the guest and exchange information through Airbnb. Due to the development of information technology industry, the company has grown fast since its founding in 2008. There are now over 150 million users worldwide and have hosted over 400 million travelers (Airbnb, 2015). Airbnb is another example of both network and crowd-based consumer power, as people share satisfactions and reviews and make decisions that would impact on both hosts and guests' reputations.

The business utilizes the concept of sharing economy in the dual environment as the process of renting out one's places, picking and booking a place to stay. All those processes are done on the online platform, but the final service will be delivered in the physical form of lodging place. One of the most important keys contributed to the business's success story focuses on customer relationship through online channels such as customer engagement, satisfaction, and feedback. They are being taken care of seriously in almost real-time manners. Airbnb has been able to convey their essence through detailed and effective social commerce. They are not only continuously reaching out to their customers' overall online community platforms, but Airbnb also shapes the way each of the platforms will interact. While they use Instagram to inspire travelers with the application's picturesque nature, Twitter is for spontaneous user communication and crowd reaction. User-generated content strategy has resulted in more than 1.8 million posts tagging with #airbnb on Instagram. The company's investment in social campaigns like "one less stranger" in 2015 and "we accept" in 2017 also has a major influence on customer's perception of the brand. As the campaigns that showed what the company value went viral, the customer acknowledged the brand personality and helped enhance the messages.

Airbnb has demonstrated all four of the consumer powers and captures positive effects from all of them. By employing CMSEs, Airbnb has driven up the demand by enticing customers with endless possibilities of new affordable traveling experiences; relying on user-generated contents, swift customer feedback, and with great uses of information sharing system, the business has established a crowd-based network of 150 million users globally.

Discussions

Marketers and businesses worldwide need to understand and utilize our CPSC framework so they can thrive in the new age and reach maximum consumers. Our research can provide guidance in this task by explaining the different types of consumer power and how they are related to social commerce, describing why firm management will need new marketing strategy and actions in this new era, explain that consumers have new power in the CPSC marketing age, and why companies must realize this with their strategic marketing. Business managers can use this information to get ahead of competitors.

The Uber case study confirms that Proposition 1, which questions if dual environments affect consumers purchase and decision-making process, is true. In this research, we were able to analyze the Uber Company, and we found that this type of business is inserted in the sharing economy model, which allows consumers to connect with each other in both virtual and physical environments. More than that, the key to the sharing economy is trust: consumers and providers must trust each other to reach positive experiences. Since Uber makes use of dual environments, by offering consumers the opportunity to order a ride online and physically experience their purchase, integrating virtual and physical environments, this can affect consumers purchase if they do not trust that the service, they receive will be worth it.

In addition, Uber allows riders to rate the drivers and the service in general. In other words, consumers can share their experience and level of satisfaction online, so other consumers are able to check how many points a driver has and decide to continue the purchase with that driver or not, based on the reviews. Therefore, this makes Proposition 2 true as well, because it confirms that consumer power increases social commerce, and that both phenomena are directly related. For instance, if a consumer rates his/her driver very well, this driver might have a trustful image on the Uber app. On the other hand, if this driver does not receive a good review, it will impact on his/her future rides in a negative way.

The Pinterest case study confirms that Proposition 3 is also true. As social commerce increases, consumer power increases as well as businesses profits through better brand management in dual CMSEs. It shows that social commerce can influence the users to go to the store and buy items they liked on the web. The reverse showrooming practice happens with a considerable part of Pinterest users, giving managers and marketers the chance to work on their customers online and improve sales by taking advantage of their profile characteristics.

The Airbnb case study confirms that Propositions 1 and 2 are valid. Throughout the research, we were able to analyze the Airbnb Company, and we found that this type of business is interjected in the sharing economy model that allows consumers to connect with each other in both virtual and physical environments. Airbnb also allows hosts and guests to share information about each other such as experiences, satisfactions, photographs, descriptions, and even personal information. The Airbnb consumers can check the reviews and ratings before deciding to purchase with the hosts.

Conclusion

This research was helpful for us to understand the relation between consumer power, social commerce, and brand management in dual environments, as well as how consumers behave in both virtual and physical environments through the decision-making process. We could analyze how consumer power impacts social commerce in CMSEs, then find that it is important for marketers to understand how to make proper use of social commerce to create a better network approach to brand management, with the intention of increasing businesses profits. In addition, we were able to develop our own conceptual framework, which is Consumer Power-based Social Commerce. Our CPSC framework can be helpful for different types of businesses in future, although it is relevant to remember that each step of the framework must receive equal amount of attention, since the concepts are integrated. Finally, we were able to find answers for the three propositions that guided our project, all of them being true, and discuss the information studied during this research.

References

Airbnb. (2015, February 27). *Creating #OneLessStranger: Stories of belonging*. Retrieved from https://blog.atairbnb.com/creating-onelessstranger-stories-belonging/

Banerjee, S., & Longstreet, P. (2016). Mind in eBay, body in Macy's: Dual consciousness of virtuo-physical consumers and implications for marketers. *Journal of Research in Interactive Marketing, 10*(4), 288–304.

Birkner, C. (2015, April). *The Goldfish Conundrum*. Retrieved from https://www.ama.org/publications/MarketingNews/Pages/the-goldfish-conundrum.aspx

Brockman, B. K., Rawlston, M. E., Jones, M. A., & Halstead, D. (2010). An exploratory model of interpersonal cohesiveness in new product development teams. *Journal of Product Innovation Management, 27*, 201–219.

Davis, B., & Hillier, L. (2019, May 7). *10 examples of great Airbnb marketing creative*. Retrieved from https://econsultancy.com/10-examples-of-great-airbnb-marketing-creative/

Edelman, B. G., & Luca, M. (2012). *Digital discrimination: The case of Airbnb.com* (Harvard Business School NOM Unit Working Paper, no. 14-054).

Edelman, B. G., & Luca, M. (2014). *Digital discrimination: The case of Airbnb.com* (Harvard Business School NOM Unit Working Paper, no. 14-054).

Fiegerman, S. (2019). Pinterest hits 300 million monthly users and its stock is soaring. *CNN*. Retrieved from https://cnn.com

Frederick, A., & Barbara, M. (2016). *Uber and the sharing economy: Global market expansion and reception*. ERB Institute - University of Michigan. WDI Publishing.

Gensler, S., Völckner, F., Liu-Thompkins, Y., & Wiertz, C. (2013). Managing brands in the social media environment. *Journal of Interactive Marketing, 27*(4), 242–256.

Glaser, B. G., & Strauss, A. L. (1965). Discovery of substantive theory: A basic strategy underlying qualitative research. *American Behavioral Scientist, 8*(6), 5–12.

Griffith, D. A., Lee, H. S., & Yalcinkaya, G. (2022). The use of social media and the prevalence of depression: a multi-country examination of value co-creation and consumer well-being. *International Marketing Review*.

Hoffman, D. L., Novak, T. P., & Stein, R. (2013). The digital consumer. In *The Routledge companion to digital consumption* (pp. 28–38).

Kane, G. C. (2009). It's a network, not an encyclopedia: A social network perspective on wikipedia collaboration. In *Academy of Management proceedings* (Vol. 2009, No. 1, pp. 1–6).

Keller, K. L. (1993). Conceptualizing, measuring, and managing customer-based brand equity. *Journal of Marketing, 57*(1), 1–22.

Kessler, A. (2014). Brian Chesky: The 'sharing economy' and its enemies. *The Wall Street Journal.*

Kuksov, D., Shachar, R., & Wang, K. (2013). Advertising and consumers' communications. *Marketing Science, 32*(2), 294–309.

Kumar, V., Bhaskaran, V., Mirchandani, R., & Shah, M. (2013). Creating a measurable social media marketing strategy: Increasing the value and Roi of intangibles and tangibles for Hokey Pokey. *Marketing Science, 32*(2), 194–212.

Labrecque, L. I., vor dem Esche, J., Mathwick, C., Novak, T. P., & Hofacker, C. F. (2013). Consumer power: Evolution in the digital age. *Journal of Interactive Marketing, 27*(4), 257–269.

Liu, Y.; Yang, J., & Liu, M. (2008, July). Recognition of QR Code with mobile phones. In *2008 Chinese control and decision conference* (pp. 203–206). IEEE.

Lowry, B. (2019, August 7). *Eric Toda, Airbnb's head of social marketing: "Stories are everything".* Retrieved from https://www.curalate.com/blog/eric-toda-airbnb/

Mabert, V. A., Muth, J. F., & Schmenner, R. W. (1992). Collapsing new product development times: Six case studies. *Journal of Product Innovation Management, 9*, 200–212.

Malthouse C., Haenlein, M., Skiera, B., Wege, E., & Zhang, M. (2013). Managing customer relationships in the social media era: Introducing the social CRM house. *Journal of Interactive Marketing, 27*(4), 270–280 (this issue).

Meng, A. (2019, January 14). *What is Pinterest, and how does it work?* Retrieved from https://www.infront.com/blog/what-is-pinterest-and-how-does-it-work/

Nafees, L., Cook, C. M., Nikolov, A. N., & Stoddard, J. E. (2021). Can social media influencer (SMI) power influence consumer brand attitudes? The mediating role of perceived SMI credibility. *Digital Business, 1*(2), 100008.

Naylor, R. W., Lamberton, C. P., & West, P. M. (2012). Beyond the "like" button: The impact of mere virtual presence on brand evaluations and purchase intentions in social media settings. *Journal of Marketing, 76*(6), 105–120.

Ng, S., & Vranica, S. (2013). P&G Shifts marketing dollars to online, mobile. *The Wall Street Journal, 2.*

Saige Driver, S. (2018, September 4). *How to use Pinterest for business*. Retrieved From https://www.businessnewsdaily.com/7552-pinterest-business-guide.html

Sevitt, D., & Samuel, A. (2013). How Pinterest puts people in stores. *Harvard Business Review, 91*(7), 26–27.

Silverman, D. (2000). *Doing qualitative research*. Sage.

Yadav, M. S., De Valck, K., Hennig-Thurau, T., Hoffman, D. L., & Spann, M. (2013). Social commerce: A contingency framework for assessing marketing potential. *Journal of Interactive Marketing, 27*(4), 311–323.

Yang, J., & Mai, E. (2010a). Experiential goods with network externalities effects: an empirical study of online rating system. *Journal of Business Research, 63*(9–10), 1050–1057.

Yin, G., Wei, L., Xu, W., & Chen, M. (2014, June). Exploring heuristic cues for consumer perceptions of online reviews helpfulness: The case of Yelp.com. In *PACIS* (p. 52).

Yin, R. K. (1994). *Case study research: Design and methods*. Sage Publications.

"US Total Media Ad Spend Inches Up, Pushed by Digital - eMarketer". (2013, August 22). Retrieved from http://www.emarketer.com/Article/US-Total-Media-Ad-Spend-Inches-Up-Pushed-by-Digital/1010154

Parallel Worlds: Human-Focused Socio-Cultural Norms and Negotiations—An AI-Free Domain

CHAPTER 9

Cross-Cultural Negotiations and the Impact of Culture in a Western-Asian Context

Anne-Sophie Bacouel

Abstract Building on Gelfand and Dyer's dynamic and psycholog-ical model of culture and negotiation, this chapter sheds light on the complexity of cross-cultural negotiations by investigating the role of perceived overconfidence bias on negotiation outcomes in a Western-Asian negotiation setting. Drawing on 12 in-depth interviews with Western negotiators, it was found that overconfident Western negotiators who perceived their Asian counterparts as being overconfident engage in short-term win-lose strategies, while non-overconfident Western negotia-tors do not perceive their counterpart as overconfident and engage in win–win strategies.

Keywords Negotiation · West · Asia · Overconfidence · Negotiation bias · Cross-cultural

A.-S. Bacouel (✉)
University of St. Gallen, St. Gallen, Switzerland
e-mail: anne-sophie.bacouel@unisg.ch

© The Author(s), under exclusive license to Springer Nature Switzerland AG 2022
A. S. Arora et al. (eds.), *Managing Social Robotics and Socio-cultural Business Norms*, International Marketing and Management Research, https://doi.org/10.1007/978-3-031-04867-8_9

Introduction

In today's global marketplace, negotiations occur increasingly across national borders. Consequently, the impact of culture on negotiations has become of major interest. Extant research has started to look at the influence of culture on negotiation tactics and outcomes (Fang, 2006; Ramirez Marin et al., 2019), proximal situational conditions, and on negotiation outcomes (Voldnes et al., 2012). However, most research ignores the complex interrelatedness of cultural variations and the proximal social conditions, the negotiator state and the resulting behaviour of culturally different negotiators. Building on Gelfand and Dyer's (2000) dynamic and psychological model of culture and negotiation, the target of this study is to highlight the complexity of cultural impact on negotiations in a Western-Asian negotiation setting and to show how humans are able to achieve win–win outcomes by adapting to the unique cultural negotiation contexts they face. Negotiation is a fundamental business skill that is as inextricably linked to human emotion and psychology. While some experts are already experimenting with artificial intelligence to improve human negotiation tactics (Schulze-Horn et al., 2020) this chapter illustrates that in cross-cultural negotiations, negotiators face constant changes in sociocultural context and thus in negotiation rules. Unlike AI, humans can learn implicitly and are able to adapt quickly to changing rules.

Theoretical Background

Negotiation Processes and Outcomes

Negotiation can be defined as a mutual persuasion process between at least two parties providing arguments in an attempt to influence each other to accept their view regarding the value of an acceptable exchange rate for a negotiated object (Robbins & Judge, 2010; Maaravi et al., 2011). Negotiation research aims at understanding the processes and outcomes of negotiations by taking into account all negotiation counterparts and the negotiation's structure (Gelfand & McCusker, 2002; Hammond et al., 2001). Different strands of negotiation research can be distinguished investigating the negotiators' behaviour, including for instance motives and goals (Pruitt, 1981) and social context (Kramer & Messick, 1995).

Impact of Culture on Negotiation

Hofstede (1980) defines culture as the collective programming of the mind distinguishing one social group and its members from another. In this sense, culture comprises shared values which result in culture-specific norms and behaviours, often responding to important societal concerns (Schwartz, 1994). This implies cross-cultural activities, including negotiations, have to take into account these different cultural settings in order to become successful (Triandis, 1982). Negotiation is a communication process by which the involved parties exchange information through signs, symbols, and behaviours which might differ cross-culturally (Morris & Gelfand, 2004). Adair and Brett (2004) claim that culture affects "people's beliefs or cognitive representations of what negotiation is all about" (p. 158), and the goals and norms they have for negotiations (Brett et al., 2017).

Research has distinguished regional cultural clusters with amongst others Confucian (East Asia) region, Western European region and Anglo-Saxon region (Chhokar et al., 2008). Western European cultures emphasise individualism, including intellectual autonomy and egalitarianism but are low on hierarchy and collectivism. Anglo-Saxon cultures were found high in individualism and mastery while low in harmony and collectivism (Lewis, 2006; Hofstede, 1980) pointing to an assertive, pragmatic, and entrepreneurial orientation (Sagiv & Schwartz, 2007). The East Asian cluster is heavily influenced by Confucianism (Fang, 2006; Hofstede & Bond, 1988; Pfajfar & Malecka, 2022). Confucian-influenced countries hold a pragmatic, entrepreneurial orientation. However, this cultural cluster combines a heavy emphasis on hierarchy and collectivism. (Sagiv & Schwartz, 2007). Finally, in tight cultural systems, such as Japan and Germany, there is less variability in the perception of situational norms, and there is greater situational constraint (Gelfand et al., 2011) on the behavioural patterns that are appropriate across a wide range of situations. By contrast, in loose cultural systems, there is a greater range of behaviours that are appropriate across situations.

Gelfand and Dyer (2000) have developed a dynamic and psychological model of culture and negotiations (see Fig. 9.1) which reflects the complexity of the interdependencies between culture and various factors directly linked to the negotiator and negotiation outcomes. We suggest that this model allows for a more complex and therefore refined depiction of cultural effects in negotiation.

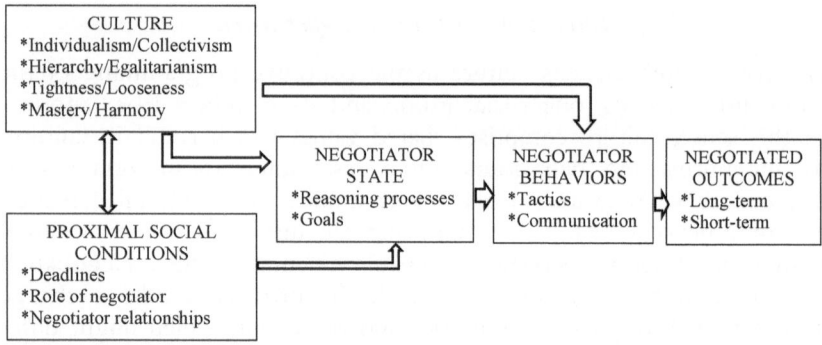

Fig. 9.1 Model of culture and negotiation (*Source* Adapted from Gelfand and Dyer [2000])

Cultural Influence on Negotiators' State

Adair and Brett (2004) suggest that culture provides the shared meaning systems which impact the way negotiators make judgements and form ways of reasoning, and the way negotiators form negotiation goals.

Culture and Reasoning Processes: Information can be processed in an analytical-rational mode or an intuitive-experiential mode (Epstein et al., 1996). Negotiators in individualistic cultures rely more on analytical-rational thinking styles and associated tactics, while collectivist cultures rely more on intuitive-experiential thinking styles and tactics (e.g. emotions). March (1988) finds evidence that Japanese negotiators tend to appeal to the feelings and goodwill of others.

Culture and Negotiation Goals: Negotiators have both relational goals and outcome goals (Olekalns & Weingart, 2001; Pinkley, 1990). Relational goals stress either trust or dominance, while outcome goals emphasise individual or joint gains. While US-Americans stress financial goals and put relational goals last, Japanese put relationships first and Latin Americans emphasise honour and prestige (Lewis, 2006). Hofstede (1980) characterises Western cultures as individualistic and independent from social affiliations, focusing on personal goals rather than social group obligations. In contrast, Asian cultures are collectivist, acting under the constraint of the norms of the social group they belong to in order to preserve harmony and face (Markus & Kitayama, 1991; Shenkar &

Ronen, 1987; Tu et al., 2021). Research has given evidence that individualistic and high mastery negotiators frame negotiations often as win-lose outcomes (Gelfand & Realo, 1999) relying on tactics such as threats or warnings. Collectivist negotiators such as Asians framed negotiations cooperatively (win–win). For instance, Graham (1993) found that Japanese restrain from commanding and threatening in negotiations indicating their concern for relationship.

Cultural Impact on Proximal Social Conditions
Proximal social conditions comprise situational conditions such as the role of the negotiator (e.g. buyer or seller), the relationship between the parties (e.g. power differentials), deadlines (e.g. time constraints) that impact negotiations (Pruitt & Carnevale, 1993). However, as social practices differ across cultures, the prevalence of certain proximal conditions might vary across cultural contexts (Markus et al., 1997). Negotiation counterparts from different cultural backgrounds may therefore expect different situational conditions as normal which can result in confusion amongst the parties.

Cultural Impact on Negotiator Behaviour
Culture and communication: Communication norms differ considerably between Asian and Western cultures with regards to context. Context is here defined as the information that surrounds an event (Hall, 1976). While Asian cultures tend to rely on high-context communication in which meaning is also communicated by the context in which words or acts are embedded, Western cultures tend to be lower context with meaning embedded in words or acts. Low-context communication is direct but does not require familiarity with contexts. The logic tends to be linear. High-context communication is indirect and requires familiarity with the (cultural) meaning conveyed by specific contexts. Therefore, the logic in high-context communication becomes more amorphic with the requirement to deduce the focus of the argument (Adair & Brett, 2004). Researchers found that low-context cultures persuade in negotiations by referring to rationality while persuasion in high-context cultures relies on emotions and relationship (Drake, 1995; Glenn et al., 1977; Johnstone, 1989). In low-context cultures, information sharing is explicit and direct, whereas in high-context cultures, information sharing is implicit and indirect (Hall, 1976). When evaluating judgemental biases (e.g. overconfidence) of the negotiation counterpart, differences in communication

styles might obfuscate the perception process and lead to erroneous conclusions.

Negotiation tactics: The prevailing negotiation goal of a negotiator as elaborated earlier impacts the negotiation tactics that will be applied. Following studies of Pruitt and Rubin (1986) and Carnevale and Pruitt (1992), it is suggested that collectivist cultures with concern for others favour maintaining relatedness with the counterparty by using indirect communication styles and avoiding strategies to achieve win–win outcomes (Gelfand & Dyer, 2000). On the contrary, individualistic cultures might engage in direct and more aggressive strategies to achieve win-lose outcomes (Tu et al., 2021).

METHODOLOGY

The study approach is built interpretative research methodology to analyse and compare the different negotiation experiences the European informants had in East Asia. Data collection was based on episodic interviewing. This method enables to gain insights about what the informant sees as significant and acceptable to communicate, as it is designed for activating interviewees to select relevant situations and to describe the contexts that make these episodes relevant to them (Flick, 2000).

Sample

The sample of this study comprised 12 European professionals in diverse industries with negotiation experience in East Asia. All informants actively took part in negotiations in East Asia with East Asian counterparts. 100% of the sample was male and the average age was 50. Their negotiation experience related to at least one East Asian country. 58% of the interviewees held a buyer role during negotiations. Interviews were conducted over Skype, audio-recorded, and transcribed. The interviews lasted around 45–60 minutes and were conducted in the informants' native language. Anonymity was guaranteed. The interview consisted of questions on the informants' personal negotiation experiences.

Analysis

The analysis was conducted in three interrelated steps. First, all the interview transcripts were thoroughly read to get an overview of the recurring themes. Second, the data was structured building on an adaptation of Gelfand and Dyer's (2000) dynamic and psychological model of culture and negotiation (see Fig. 9.1). Narratives were coded from the interviews along the themes suggested in the presented model. The third step of the analysis encompassed working out interconnections between the coded narratives by referring to the possibly moderating influence of culture.

Findings

Impact of Culture on Proximal Social Conditions

Negotiator Relationship
Asian countries are collectivist and strongly rely on relationship and trust building. Therefore, negotiator relationship is key in the East Asian culture, and most of the interviewees adhered to that cultural element and put emphasis on a good relationship with the East Asian counterpart, even citing it as an important behaviour for them.

> I think in China you shouldn't underestimate the importance of a personal relationship. Having dinner and being more relaxed is a very important part of making the final deal. And for me, in that situation, it was a good opportunity to develop a more personal relationship with the customer. (British, seller)

Asian collectivist and high-hierarchy cultures tend to involve groups with representatives based on ascribed status into negotiations who often face single individualist and autonomous Western negotiators selected for their achieved status. Thus, it is important for Western negotiators, who are mainly autonomous in negotiations, to decode the hierarchical decision-making process when facing a group of East Asian counterparts.

> You must look for who is interested in what on the other side and then exactly try to get the right level and the people who can break a dead-lock. That usually works through hierarchy. Now you can't easily jump hierarchies, that is also dangerous, because you piss off the people at the working level. (German, negotiator)

Knowing about these cultural differences and being hierarchically higher placed or connected than the Asian negotiators seems to enhance the power position of some Western negotiators.

Deadlines
Different timeframes on the East Asian side is also a widely discussed area in the interviews. The findings here are somewhat contradicting, with some interviewees highlighting the slowness of the process due to the time spent on relationship building and complicated hierarchies on the Asian side.

> As a European, you have a huge agenda you want to go through, and you also want to reach your goals. And they [the Chinese] just say, „Yes, yes, no problem". But before that it was just chatting all the time [...]. And then you get to a point where it's like "Come to the point! I don't have time all day for things like this". (German buyer)

The narratives underline that structuring time is quite rigid in Western analytical-rational cultures, where negotiators like to proceed according to a clear timeframe. East Asians use a more intuitive-experiential mode (Epstein et al., 1996) which adapts to persons and new situations:

> [A] lot of the western negotiations are based on established formal positions while the Chinese position is much more. (French, seller)

The reliability of the East Asians in terms of respecting the deadlines set in the contract and the length of the negotiation process was evaluated differently by the informants. While some see for instance Chinese positions as everchanging (see the narrative above), others experienced that,

> If they [the Chinese] said they were going to do something, whether it was in writing or not, they did it. (French legal)

Impact of Culture on the negotiator's State

Culturally Different Reasoning Processes
All respondents mentioned the Western rational way of thinking, its translation into the negotiation processes and the mismatch with the valuation of personal relationships on the East Asian side.

> Western negotiations are very formal, include going through a lot of bullet points, extremely rational. The Chinese typically try to get around more

social settings, like wine and dining. Discussions are not only done in a formal setting. (French, seller)

These narratives confirm existing research findings, and the distinction of Asian relationship building versus Western rational deal focus (Adair & Brett, 2004) is well-recognised by all interviewees.

Negotiation Goals
Individualistic Western cultures aim at more competitive outcome goals while Asian collectivist cultures tend to favour more cooperative relational goals (Lewis, 2006) resulting in either win-lose or win–win negotiation tactics. There is evidence for the case of few Western negotiators in our sample who were aiming for a "win-lose" outcome in their negotiations. However, all the other informants agreed in reaching a "win–win" situation to preserve a long-term relationship. However, those engaging in win-lose games do not seem to perceive incompatibility with long-term relationship as long as the negotiation outcomes are evaluated as fair.

Cultural Impact on Negotiator Behaviour

Communication
Interviewees experienced communication issues, not only due to the language barrier, but also due to the differing communication styles used by their counterparts.

> Japan is a very high context culture and Germany is very low context. In Japanese negotiations, it's very important to understand what is not being said. In Germany you put all the things on the table, and it will be a tough negotiation. (German, buyer)

This shows that collectivist/individualist cultural conceptualisations determine negotiation behaviour and communication: High context / harmony versus low context/mastery (Hall, 1976). However, this is seen as a major obstacle by Western negotiators who recognise the cultural feature and put effort in developing strategies to advance negotiation processes while still respecting the Asian communication style:

> In the Chinese culture, there is no 'no' and so it is important to let the negotiation partner summarise important points in the negotiation

and thus get an understanding if he understood it as I just explained it. (German, buyer)

Negotiator Tactics
In line with previously discussed findings, East Asian negotiators were typically found to be encouraging harmony in negotiations, emphasising good relationships with their Western counterparts, and accommodating their needs.

> So, they [the Chinese] would always try to see what you are like, what's your character and then they would start the negotiation, to maximise the possibility of agreeing and meet your motivations. (French, seller)

Most Western negotiators of the sample recognised this difference in negotiation behaviour and most adapted to these mainly as a means to an end to reach their desired results.

DISCUSSION AND IMPLICATIONS

The aim of this study is to examine the impact of culture in cross-national negotiations and on long-term and short-term relationship between the negotiators. The presented findings demonstrate that this investigation calls for a complex analysis of various interrelated impact factors. Within a negotiation process, proximal social conditions and the negotiators' psychological states influence their negotiation behaviours and accordingly the negotiation outcome. These factors are influenced by the cultural setting in which the negotiation is embedded (Gelfand & Dyer, 2000). Gelfand and Dyer (2000) state that intercultural negotiation needs to be understood as a dynamic and interrelated system of culture, proximal social conditions, psychological states, negotiation behaviour, and outcomes. Extant studies (e.g. Fang, 2006; Kumar, 1999) have so far neglected the complexity of these interrelations. Building on Gelfand and Dyer's (2000) model, the findings in this study clearly confirm the complexity of intercultural negotiations and challenge static models which attribute cultural features of negotiation partners with a static culture-specific negotiation behaviour and outcome.

In the context of globalisation of trade and the growing importance of Asian markets and sourcing destinations for Western economies, not only intercultural negotiations in general but also their frequency have significantly increased (Fang, 2006). Western companies trading with Asia considered negotiation strategy an important success factor for trading relationships (Fang, 2006). Intercultural negotiations are often assumed to be more difficult and associated with lower outcomes than intracultural negotiations. Awareness about this complexity of cross-cultural negotiations may help managers to better design cross-cultural training or negotiation programmes.

CONCLUSION

Our findings illustrate, based on the example of Western-Asian negotiation contexts, the high complexity of intercultural negotiations and allow a more refined picture of the interrelatedness of various factors that influence intercultural negotiation outcomes and could be a starting point for future research to investigate these issues more in-depth.

REFERENCES

Adair, W. L., & Brett, J. M. (2004). Culture and negotiation processes. In M. J. Gelfand & J. M. Brett (Eds.), *The handbook of negotiation and culture* (pp. 158–176). Standford University Press.

Brett, J. M., Gunia, B. C., & Teucher, B. M. (2017). Culture and negotiation strategy: A framework for future research. *Academy of Management Perspective, 31*(4), 288–308.

Carnevale, P. J., & Pruitt, D. G. (1992). Negotiation and mediation. *Annual Review of Psychology, 43*, 531–582.

Chhokar, J. S., Brodbeck, F. C., & House, R. J. (2008). *Culture and leadership across the world: The GLOBE Book of in-depth studies of 25 societies.* Lawrence Erlbaum Associates.

Drake, L. (1995). Negotiation styles in intercultural communication. *International Journal of Conflict Management, 6*, 72–90.

Epstein, S., Pacini, R., Denes-Raj, V., & Heier, H. (1996). Individual differences in intuitive-experimental and analytical-rational thinking styles. *Journal of Personality and Social Psychology, 71*(2), 390–405.

Fang, T. (2006). Negotiation: The Chinese style. *Journal of Business and Industrial Marketing, 21*(1), 50–60.

Flick, U. (2000). Episodic interviewing. In M. W. Bauer & G. D. Gaskell (Eds.), *Qualitative researching with text, image and sound* (pp. 75–92). Sage.

Gelfand, M. J., & Dyer, N. (2000). A cultural perspective on negotiation: Progress, pitfalls, and prospects. *Applied Psychology, 49*(1), 62–99.

Gelfand, M. J., et al. (2011). Differences between tight and loose cultures: A 33-nation study. *Science, 332.*

Gelfand, M. J., & McCusker, C. (2002). Metaphor and the cultural construction of negotiation: A paradigm for theory and research. In M. Gannon & K. L. Newman (Eds.), *Handbook of cross-cultural management* (pp. 292–314). Blackwell.

Gelfand, M. J., & Realo, A. (1999). Individualist—Collectivism and accountability in intergroup negotiations. *Journal of Applied Psychology, 84,* 721–736.

Glenn, E. S., Witmeyer, D., & Stevenson, K. A. (1977). Cultural styles of persuasion. *International Journal of Intercultural Relations, 1*(3), 52–66.

Graham, J. L. (1993). The Japanese negotiation style: Characteristics of a distinct approach. *Negotiation Journal, 9*(2), 123–140.

Hall, E. T. (1976). *Beyond culture.* Anchor.

Hammond, S., Keeney, R. L., & Raiffa, H. (2001). The hidden traps in decision making. *Harvard Business Review on Decision Making.*

Hofstede, G. (1980). *Culture's consequences: International differences in work-related values.* Sage.

Hofstede, G., & Bond, M. H. (1988). The Confucius connection: From cultural roots to economic growth. *Organizational Dynamics, 16*(4), 5–21.

Johnstone, B. (1989). Linguistic strategies and cultural styles for persuasive discourse. In S. Ting-Toomey & F. Korzenny (Eds.), *Language, communication and culture* (pp. 139–156). Newbury Park, CA: Sage.

Kramer, R. M., & Messick, D. M. (1995). *Negotiation as a social process: New trends in theory and research.* Sage.

Kumar, R. (1999). Communicative conflict in intercultural negotiations: The case of American and Japanese business negotiations. *International Negotiations, 4*(1), 63–78.

Lewis, R. D. (2006). *When cultures collide: Leading across cultures* (3rd ed.). Nicholas Brealey International.

Maaravi, Y., Ganzach, Y., & Pazy, A. (2011). Negotiation as a form of persuasion: Arguments in first offers. *Journal of Personality and Social Psychology, 101*(2), 245–255.

March, R. M. (1988). *The Japanese negotiator: Subtlety and strategy beyond Western logic.* Kodansha International.

Markus, H. R., & Kitayama, S. (1991). Culture and the self: Implications for cognition, emotion, and motivation. *Psychological Review, 98,* 224–253.

Markus, H. R., Kitayama, S., & Heiman, R. J. (1997). Culture and "basic" psychological principles. In E. T. Higgins & A. W. Kruglanski (Eds.), *Social psychology: Handbook of basic principles.* Guilford Press.

Morris, M. W., & Gelfand, M. J. (2004). Cultural differences and cognitive dynamics: Expanding the cognitive perspective on negotiation. In M.

J. Gelfand & J. M. Brett (Eds.), *The handbook of negotiation and culture*. Standford University Press.

Olekalns, M., & Weingart, L. R. (2001). *Negotiators talk: An analysis of communication processes in negotiation* (Working Paper No. 2001–19). Melbourne, Australia: Melbourne Business School.

Pfajfar, G., & Malecka, A. (2022). Evaluating the role of Confucian virtues in Chinese negotiation strategies using a Yin Yang cultural perspective. *European Journal of International Management, 17*(2–3), 290–323.

Pinkley, R. (1990). Dimensions of conflict frame: Disputant interpretations of conflict. *Journal of Applied Psychology, 75*, 117–126.

Pruitt, D. G. (1981). *Negotiation behavior*. Academic Press.

Pruitt, D. G., & Carnevale, P. J. (1993). *Negotiation in social conflict*. Open University Press.

Pruitt, D. G., & Rubin, J. Z. (1986). *Social conflict: Escalation, stalemate, and settlement*. McGraw-Hill.

Ramirez Marin, J., Olekalns, M., & Adair, W. (2019). Normatively speaking: Do cultural norms influence negotiation, conflict management, and communication? *Negotiation and Conflict Management Research, 12*(2), 146–160.

Robbins, S. P., & Judge, T. A. (2010). *Essentials of organizational behavior*. Pearson.

Sagiv, L., & Schwartz, S. H. (2007). Cultural Values in Organizations: Insights for Europe. *European Journal of International Management, 1*(3), 176–190.

Schulze-Horn, I., Hueren, S., Scheffler, P., & Schiele, H. (2020). Artificial intelligence in purchasing: Facilitating mechanism design-based negotiations. *Applied Artificial Intelligence. An International Journal, 34*(8), 618–642.

Schwartz, S. H. (1994). Beyond individualism/collectivism: New cultural dimensions of values. In U. Kim, H. C. Triandis, C. Kagitcibasi, S.-C. Choi, & G. Yoon (Eds.), *Individualism and collectivism: Theory, method and applications* (pp. 85–122). Sage.

Shenkar, O., & Ronen, S. (1987). The cultural context of negotiations: The Implications of Chinese interpersonal norms. *The Journal of Applied Behavioral Science, 23*(2), 263–275.

Triandis, H. C. (1982). Dimensions of cultural variation as parameters of organizational theories. *International Studies of Management and Organization, 12*, 129–169.

Tu, Y.-T., Lin, C.-Y., Moslehpour, M., & Qiu, R. (2021). An intercultural comparison of negotiation styles between Taiwan and the United States. *Academy of Strategic Management Journal, 20*, Special Issue 6, 1–18.

Voldnes, G., Grønhaug, K., & Nilssen, F. (2012). Satisfaction in cross-cultural buyer-seller relationships—Influence of cultural differences. *Industrial Marketing Management, 41*, 1081–1093.

CHAPTER 10

The Effects of Work Culture on Women's Career Advancement: A Comparison Between the Netherlands and the United States

Ashley Hoek and Fabian Bernhard

Abstract This chapter describes seven distinct cultural workplace differences between the United States and the Netherlands and discusses practical implications. These differences are directness in language, work life mentality, levels of collaboration, consensus, part-time work, maternity and paternity leave, and how success is defined in the United States versus the Netherlands. Interviews with six expatriates offer additional insights into these differences. The chapter then looks at the effects on women and their career potential and suggests that the Netherlands can

A. Hoek · F. Bernhard (✉)
EDHEC Business School, Paris, France
e-mail: Fabian.Bernhard@edhec.edu

A. Hoek
e-mail: Ashley.Hoek@edhec.com

143

A. S. Arora et al. (eds.), *Managing Social Robotics and Socio-cultural Business Norms*, International Marketing and Management Research, https://doi.org/10.1007/978-3-031-04867-8_10

offer a more conducive workplace for women than the United States. However, once a woman enters motherhood, our research indicates a shift in Dutch ideology. A mother's desire to work full-time, part-time, or stay at home can influence which of the two countries better supports her career progression. The United States' work culture norms better align in the case of a mother that would like to continue to work full-time.

Keywords Netherlands · United States · Workplace · Culture · Women · Career · Comparative study

Introduction

This chapter compares workplace cultures between the United States and the Netherlands to propose which country is more favorable for women's career advancements. The traditional figure of someone who moves abroad for work or works internationally is male, with women more often relocating for their partner (Expat Insider, 2018). This can lead to the narratives about places and their opportunities to be skewed and male focused. While digital technologies, including AI, affect our culture and the realities of work, including the potential career advancement of women, the focus of this chapter is AI free and completely, human-focused domain. The research presented in this chapter aims to take on the female perspective on how two distinct countries' (United States versus Netherlands) workplace cultures fare for women who are looking to continue and advance their careers. When writing this chapter, we aimed at minimizing stereotypical considerations and relied on empirical findings found in existing works and interviewed experts.

Through literature and interviews seven distinct differences in workplace cultures were identified and described. These differences were examined under the lens of how they affect women's careers; in part, how these differences allow women to fit into the workplace and, in the larger sense, if they allow women equal opportunities for career advancement as their male peers. Based on the findings in the literature and interviews on these seven distinct cultural aspects between the United States and the Netherlands, four key findings were made. This chapter will explore research on Dutch and American workplace cultures, describe key traits that are different, and analyze what the literature says about how these traits play into women's career advancement.

THEORETICAL BACKGROUND

General Country Comparison

The World Economic Forum publishes statistics on country workforces and includes breakdowns by gender where relevant. The Netherlands ranked 38 out of 153 countries with a score of 0.736 on a scale of 0–1 on the Global Gender Gap Index. The Index is made up of five subsections: economic participation and opportunity, educational attainment, health and survival, and political empowerment. The United States ranked 53 overall with a score of 0.724. The Netherlands and the United States are equal when considering the value of the labor force (both countries score 0.47 on a scale of 1.0) and the evaluation on advancement of women to leadership roles (both achieve high scores with 5.25 on a scale of 1–7). While broadly the same, there are some notable differences in relation to unemployment percentages. A higher percentage of women is unemployed in the Netherlands, while the percentage of unemployed men is higher in the United States. Gender parity in technology fields is higher in the United States and the gender pay gap is larger in the United States. A key notable difference between the two countries can be found in part-time work: The Netherlands has drastically higher percentages of part-time workers of both genders. In the Netherlands, 40% of men in the labor force are part-time workers versus 16% in the United States. These figures rise to 75% of women in the Netherlands, but only 28% in the United States women belong to the part-time workforce.

It is important to note that this chapter will mainly focus on participation in the workforce and opportunity inequalities, not on education. As both nations have high rankings in education equality, most gender disparities in the workplace do not seem to primarily stem from inequalities in the ability to obtain an equal level of education or differences in the quality of education in these two nations.

Dutch Work Culture

As illustrated in Fig. 10.1 of the main cultural workplace indicators, the largest differences between the Netherlands and the United States

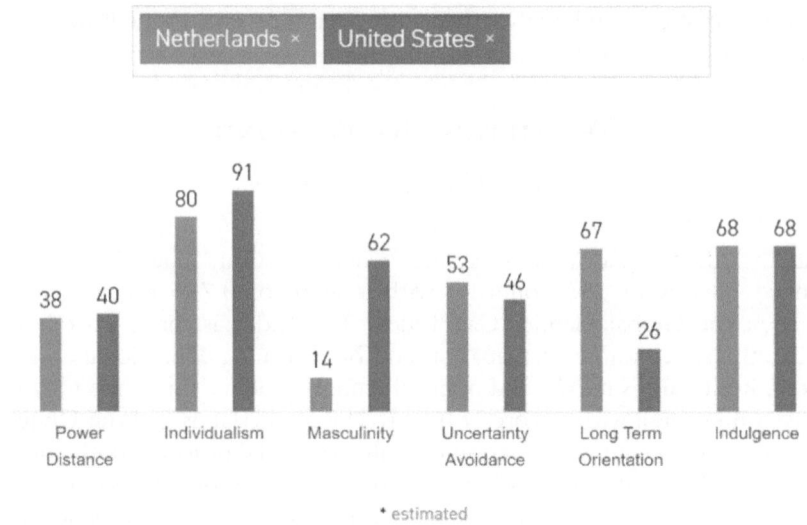

* estimated

Fig. 10.1   Country comparison (Hofstede et al., 2010)

are in the masculinity category and long-term orientation category. The real difference between a masculine and feminine society is the driving motivational forces. According to the "Hofstede" studies, the Netherlands' low score of 14 in masculinity was due to a culture where "the dominant values in society are caring for others and quality of life" (Hofstede, 2010). In a masculine society what people strive for is achievement, showing competence, and wanting to be the best at what they do (Hofstede, 2010). In a feminine society, this is less important, and the driving force is the enjoyment in what you do (Hofstede, 2010). The emphasis on these societal values plays a big aspect in work life. For example, in the Netherlands, a person is expected to enjoy his or her work and maintain a healthy personal life. Working long hours or working on the weekends is not respected, only 0.5% of the Dutch population work over 50 hours per week (International Business Guides: The United States, 2020).

Consensus in the workplace also aligns with a feminine society and is very important in Dutch culture (International Business Guides: Netherlands, 2020). The desire for consensus explains why the Dutch are fond of meetings (Jong, 2018). Dutch meetings are precise and punctual, and they follow a strict agenda that does not include small talk at the start (Jong, 2018). Every employee in a meeting is encouraged to speak and give opinions. As indicated by the lower score in Power Distance (see Fig. 10.1), any employee in a meeting is equally invited to share his or her opinions and thoughts (Mobility, 2020). When a decision is made, it is vital that each employee agrees with the idea to move forward. Often the decision-making process is slower in the Netherlands since all ideas are open to discussion and time will be spent forming agreement among the team (Jong, 2018). However, once there is consensus and a plan, progress can move forward quickly.

Finally, long-term orientation is the other area where the United States and the Netherlands largely differ. The Netherlands, with a high score of 67 out of 100, is considered a pragmatic society (Hofstede, 2010). They value thriftiness and education and see those values to prepare for the future. However, the Netherlands is also much more adaptable with their traditions, and despite a high score in long-term orientation, more quickly adapt traditions to the context of modern times. The Netherlands' early adoption of maternity leave and part-time work options is a good example of this. In contrast, the United States rarely offers such options. The United States has a relatively low score on long-term orientation because it is more hesitant to change standing norms and traditions. They are also more focused on the current moment (Hofstede, 2010). This can be seen in US business practices with standard quarterly reports on earnings and profits, rather than on an annual basis. Many high-level, C-suite positions have compensation tied to earnings rather than a set salary (Big Red Car, 2013). This causes the C-suite to push for higher earnings in the United States; a drive that is reflected in Hofstede's study with both a lower score in Uncertainty Avoidance and a significantly lower score on long-term orientation than the Netherlands. The coupling of these two aspects show how Americans live more in the present and focus on immediate gains rather than a long-term orientation.

The Dutch thriftiness explains why they are less focused on appearances than Americans. This comes across in speech, dress, and material possessions (International Business Guides: Netherlands, 2020). The Dutch preference is to spend money on leisure time and activities rather than developing a materialistic image based on possessions and appearance. This is also reflected in a more casual workplace dress code (International Business Guides: Netherlands, 2020).

Another way the Dutch demonstrate that they are less elaborate than Americans is by their directness and blunt nature of speaking. This is a dominant trait for which the Dutch are known for and is quite common in the workplace (Mobility, 2020). This directness is not interpreted as rude, but rather honest and pointed. Flattery and embellishments in speech go against the Dutch's modest and humble manner (Kemp, 2018). Since the Dutch are used to a direct style, misinterpretations occur when an individual talks around a point or uses sarcasm rather than say explicitly what he or she means (Evanson, 2016). While Americans can also be considered direct, they are more limited by political correctness and what is considered polite conversation (Conway et al., 2017). Although, according to authors Conway, Repke & Houck, some Americans are at odds with being politically correct, and many feel that it has gone too far. Conway and colleagues suggested that former United States President Trump's popularity comes in part from his lack of political correctness and his willingness to say anything (Conway et al., 2017). Many Americans feel social pressure to restrict their own communication publicly and feel unable to speak their minds (Conway et al., 2017). In contrast, one of the most important traits of a good manager in the Netherlands is honesty (Selvarajah et al., Characteristics of high performing managers in The Netherlands, 2012).

United States Work Culture

Figure 10.1 shows the United States has a high score of 62 in Hofstede's Masculine category. This greatly affects the way Americans define success when compared to the Dutch. In the United States, traditional success is seen as excelling at work, and a person's job shapes that individual's persona (Gallup, Inc.; Populace, Inc., 2019). Title and position at work provide a person with social status (Gallup, Inc.; Populace, Inc., 2019). Since exact dollar amounts and salaries are not normally disclosed, Americans tend to show their success through material objects, such as a large home, fancy car, and expensive clothing (Gallup, Inc.; Populace, Inc., 2019). The connection between salary and professional title to social status and sense of success leads the United States to having a very competitive work environment and a never-ending fight for a promotion and higher wage (Gallup, Inc.; Populace, Inc., 2019). While the Success Index study by Gallup shows that many Americans are starting to change their own ideas about success in recent years, the majority hold on to the more traditional American definition of success (Gallup, Inc.; Populace, Inc., 2019).

Success and America's "work is life" mentality stems from the "American dream" that anyone can be get ahead if they work hard enough (Barone, 2020). Low salaries and unemployment are often associated with laziness or acceptance of the mundane which is looked down upon in the United States (Rodriguez, 2017). It is common to see Americans working long hours or even having multiple jobs. The normal work week is 40 hours. This is true and standard for certain industries and job types, particularly in positions that require unskilled labor. Hourly workers must be provided overtime compensation whereas salary employees are not (U.S. Department of Labor Wage and Hour Division, 2020). However, in higher salary level positions, a 40-hour work week is often considered the bare minimum. Most employees that are paid on a salary basis are considered exempt from being paid overtime and are expected to stay until the work is done without additional compensation (U.S. Department of Labor Wage and Hour Division, 2020). The average number of hours worked a week in the United States is 38.59. By contrast, the average number of hours worked in the Netherlands is 29.30 per week (Clockify, 2020). This does not mean that the Dutch are lazy or do not work hard. The Netherlands is almost equal when comparing GDP per hours worked (Clockify, 2020). The total hours worked has remained fairly constant since around 2000 for both the United States and the Netherlands despite the advancement of smartphone and access to work without being physically at the office (Clockify, 2020). While these statistics may infer the Dutch are simply more efficient, it conflicts with another American standard: Time is money. Idle time to Americans is considered wasteful, so often the phrase "there is always something to do" will be implied to workers who have extra time on their hands or leave too early.

The American ideology of "work is life" is also evident in respect to annual leave, with US employees receiving far fewer vacation days per annum when compared with other developed countries. Contrary to the EU which offers anywhere from 20 to 28 vacation days a year (Paid Vacation Days Europe, 2021). In the United States, there is currently no legal requirement to provide vacation days and most companies offer just ten days a year on average (Hess, 2018). More vacation days are often awarded to employees as they move into higher positions and stay with the company for longer; however, it is common for managers to respond to emails and be available while on vacation (U.S. Bureau of Labor Statistics, 2018). The "work is life" mantra also drives many Americans to not use all the allocated vacation time, with 55% of all employees not utilizing their annual quota (Kim, 2019).

In addition to the "work is life" mentality, part of the American Dream emphasizes individuality and a "do-it-yourself" mentality (Hofstede, 2010). Employees are expected to be able to work independently and take initiative. It is common for employees to take on additional tasks outside their job description, or to train, volunteer, or attend events above and beyond designated working hours. Typically, the main objective here is for employees to add to their resume or gain a skillset needed to further progress their career (ET Online, 2018). Because the United States has more of an individual focus, they do not look for the consensus of the team to make decisions. Managers will often make decisions rather quickly and decisively without obtaining employee buy-in (Hofstede, 2010). However, this does not mean that Americans cannot work in a team or that they dislike teamwork. Often, working well with others is seen as a necessary means in obtaining a promotion or to be considered a good manager (Half, 2020). Recent trends, however, have shown a shift in both the idea of the aforementioned "live to work" philosophy and the general belief that profit is key (with employee satisfaction secondary). Unlike the generations before them, millennials in the United States value their vacation and personal time more, and are demanding more inclusive, employee-oriented work environments (Parmelee, 2020). In addition to valuing their personal time, millennials also think it is very important to contribute to the community (Allen, n.g.). Millennials have been a driving force for flexible work schedules and remote work. Among millennials, it is generally believed that "because of technology, they can work flexibly anytime, anyplace and that they should be evaluated on work product not how, when, or where they got it done" (Allen, n.g.). This mentality has met mixed reviews from other generations that still hold most top senior positions (Allen, n.g.). However, as millennials reach senior positions in organizations, they are in positions to make policy changes. Companies now find that they tend to have a competitive advantage in recruiting talent if they can offer their employees a better work environment and can offer a more even work life balance (Parmelee, 2020).

Difference in Part-Time Work Between the United States and the Netherlands

The United States and the Netherlands differ considerably when it comes to part-time work. The Netherlands has very strong part-time work laws that protect employees. It has one of the highest proportions of both men and women that work part-time compared to other developed nations. Of the total working population in the Netherlands, 75% of females and 40% of males work part-time (Crotti et al., 2020). These high proportions of the workforce population engaging in part-time work are said to shape the working norm in the Netherlands. "The growth of part-time work has led to a gradual decline in the adherence to the work obligation norm (Wielers & Raven, Part-Time Work and Work Norms in the Netherlands, 2013)." Work obligation here is defined as a feeling of social expectation to engage in some form of paid or unpaid work. This result is a tendency for the Dutch to feel less pressured by society to work. A person's desire to contribute to something greater is often fulfilled by charity and volunteering at a foundation as opposed to a job. Work in the Netherlands is therefore much more driven by financial need and a "work to live" mentality—a direct contrast to the American "live to work" approach.

A large majority of the female part-time working population stems from mothers that work part-time hours to help with family needs and obligations. It is also common for mothers to choose not to return to full-time work even after family obligations have been met (Wielers & Raven, Part-Time Work and Work Norms in the Netherlands, 2013). This is reflected by the Dutch feminine score and value on personal leisure time presented by Hofstede (see Fig. 10.1). Culturally, having the time to dedicate to family holds much more importance than work life for mothers. In addition, working part-time does seem to have a negative effect on career progression as seen in (Chung & Lippe, Flexible Working, Work–Life Balance, and Gender Equality: Introduction, 2018). In most cases, there is an upper limit to career advancement for an employee of either gender that indefinitely works part-time (Breeschoten & Evertsson, 2019). With more than three out of four women in the Dutch workforce working part-time, career progress may be limited with a negative impact on the potential for promotion. This explains the lower ranking on the World Economic Forum Global Gender Gap Report for Economic Participation and Opportunity when compared to the United States. The Netherlands'

Global Gender Gap Index score on Economic Participation and Opportunity is 60 out of 153 countries whereas the United States scored 26. According to authors Benschop, Brink, Doorewaard, and Leenders, one of the reasons part-time work is so prevalent in the Netherlands is because of the organizational and social pressure for mothers to look after the children. Once a woman becomes a mother, she is expected to follow the ideology of family first, work second. The norm for mothers is to work three days a week. "The idea of selling her children short by spending more time at work makes this participant feel guilty...Friends who do not adhere to this ideology are condemned" (Benschop et al., 2013).

By stark contrast, 27.99% of the female workforce in the United States, and 16.46% of male workforce, are part-timer employees. One explanation for these lower percentages, when contrasted with the Netherlands, is the fact that American work culture is still tied with identity and success. Working part-time does not show that you are committed to the organization. According to (U.S. Bureau of Labor Statistics, 2020), in 2019 about 27% of part-time workers were either students or retirees collecting retirement benefits. Individuals that are retired but still hold part-time work either desire to continue their contribution to the workforce or have a financial need to work.

Another factor contributing to the lower numbers of part-time workers in the United States is a high cost of living that necessitated both partners to work full-time. Part-time work, in many cases, may not be enough to support a family (Ross & Bateman, 2019). In the United States, even when part-time work is temporary, it is often difficult for either male or female employees to advance at a normal rate upon return (Dikkers et al., 2010). This is more the case for women, who work part-time to meet family obligations. Employers assume that a female employee that is willing to work part-time for her family will always prioritize family life overwork (Chung & Lippe, Flexible Working, Work–Life Balance, and Gender Equality: Introduction, 2018). There is an expectation that anyone working part-time will halt their career advancement until full-time work is resumed: "The widespread organizational norm that it is impossible to combine part-time work and managerial responsibilities is accepted as a matter of course. This implies that upward career mobility is not simply seen as a negative ambition, but also as an unattainable goal for part-timers" (Benschop et al., 2013).

In the Netherlands, stunted career progression disproportionately affects mothers who make up most women that work part-time. The mix of social pressure to stay home with the children and the desire to take care of the family is a big driver for women to knowingly choose part-time work despite understanding the effect on her career. "They [women] relate negatively to ambition as 'upward career mobility' and they agree that working part-time – self-evidently – hinders upward career mobility" (Benschop et al., 2013).

Difference in Maternity Leave Between the United States and the Netherlands

The United States does not have federally mandated paid maternity or paternity leave. It does have the Family and Medical Leave Act (FMLA), which requires employers to give up to twelve weeks of unpaid leave for "the birth and care of the newborn child of an employee, for placement with the employee of a child for adoption or foster care, to care for an immediate family member (i.e., spouse, child, or parent) with a serious health condition, or to take medical leave when the employee is unable to work because of a serious health condition (U.S. Department of Labor, n.d.)." This usually cannot be combined with sick leave and companies can require sick leave to be deducted from the 12-week period (U.S. Department of Labor, n.d.). FMLA only applies if the employee has "been at their current job for at least 12 months, worked 1,250 hours in the previous year and are employed by a company with at least 50 employees within a given 75-mile radius (U.S. Department of Labor, n.d.)." If FMLA is not an option and the employer does not offer maternity leave, new mothers are often allowed to use sick leave and vacation days together to get paid time off to take care of a newborn, but this is at the discretion of the company (Zintl, n.d.). This is often limited to the mother whereas fathers can only use their allotted vacation days (Dennison, n.d.). New mothers and sometimes expecting mothers have the option to take leave without pay if they can afford to do so, however, this is also at the discretion of the company (Zintl, n.d.).

Taking an extended break for family reasons is highly criticized in the United States (Dennison, n.d.). It is often perceived as putting family priorities ahead of work and has a negative correlation with future promotions (Johns, 2013). This results in women with families struggling to advance their careers at a comparable rate to their male peers in the United States (Johns, 2013). Even perceptions on telework in the United States have a similar negative connotation for females compared to males.

Women that telework and have families are seen to do so to be able to better care for their families while men do so to be more dedicated to work. The perception is that men who telework do not have office distractions or take time with a commute and are thus better able to focus on work whether the male employee has a family (Chung & Lippe, 2018).

The Netherlands, on the other hand, offers both maternity and paternity benefits which are not limited to new parents. A Dutch woman can receive up to sixteen weeks of maternity leave with 100% of earnings in a twelve-month period. In most cases it is required that the expecting mother start the maternity leave four weeks before the baby is due and immediately take the remaining twelve weeks. This is the standard for a normal full-term pregnancy (Netherlands Enterprise Agency, RVO, 2020). Partners are allowed one week of paternity leave at 100% of their salary paid by the employer. They are also eligible for an additional six weeks of leave at 70% of their salary paid, which is paid by the government and can be taken at any point in the first six months of the child's birth. Employers have the right to deny this but must provide valid justification for doing so and must make alternative arrangements with the employee (Lalor, 2020). Many companies will offer even more leave for partners, above and beyond the basic legal requirements. As Lalor states, "ING gives fathers one month of leave and an additional two months unpaid. The Hague municipality allows its employees to have 6.5 weeks off with full pay when they become parents" (Lalor, 2020).

The Netherlands also offers unpaid parental leave which can be used by both partners. It amounts to 26 times the hours worked and can be used anytime until the child turns eight years old. For an employee that averages a 40-hour work week, that equals approximately twenty-six weeks of unpaid leave over eight years (Steck, 2020).

There are, however, mixed ideas about whether maternity leave is good for women's career advancement. Maternity leave allows a woman a pause in her career with the intention that she will be able to return to work more seamlessly and thus not have to choose between having a family and working. Countries that offer maternity leave by law protect the mother, however, it is hard to prove whether these countries hold an underlying bias against promoting women of childbearing age for the fear that they will take extended time off (C., 2017). A trend has been observed that suggests there is a temporary lag in promotions for women who have taken maternity leave versus other employees that have taken a similar length leave for other reasons (Thomas, 2016).

Different Aspects of Work Culture Effect on Work Advancement

Directness: Research indicates that females in the workplace are less direct in their language than men and more polite and reserved. As Johns explains, "women's typical communication style is more warm, less directed, and more mitigated than a typical man. This style of communication can lower perceptions about women's abilities" (Johns, 2013).

This is illustrated in Table 10.1 where we see a stereotypical general representation of male working styles and female working styles. Typical male directness aligns more with Dutch work culture than a female's persuasive approach. Additional research comparing Dutch women and American women could be done to evaluate if these stereotypes hold and what effect they may have within the two different cultural contexts. As illustrated in Table 10.1 most of the stereotypically female traits are more aligned with Dutch work culture than they are with American work culture. Women are depicted as being more relationship oriented rather

Table 10.1 A balance of both masculine and feminine strengths (Turner, 2012)

Workplace area	Male approach	Female approach
Structure—how they structure organizations, teams, and work	Hierarchy—levels and status matter; power and information flow top to bottom	Network—Relationships matter more than status; roles overlap; power and information are shared
Work Focus—what is forefront when they make a decision	Goal Focus—keeps focus on the goal/result; pushes aside distracting ideas	Process Focus—focuses also on the process to get the result; gathers multiple inputs; considers related issues
Influence—how they get others to do what they think is important	Command—gives orders; tells; is direct, clear	Persuade—makes requests, asks; is indirect, polite
Work Style—what energizes them when working with others	Compete—sees work as a game to be won; coming out on top is key	Collaborate—gets results through relationships, involvement and sharing power
Conflict—how they resolve disputes	Direct—confronts directly; facts are most important; seeks closure; conflict isn't personal	Indirect—uses indirect approaches; emotions and facts both matter; holds on to hurt feelings

than results driven like their male peers—a valued characteristic in Dutch workplace culture.

Consensus and collaborative work environment: While Turner makes it clear that these are stereotypes, it does not consider the pressure women experience in the workplace to act more like their male peers (Cassard & Hamel, 2008). Women in the United States might have some social pressure to act more traditionally male, whereas female employees in the Netherlands would have pressure to act more traditionally female. This distinction could play into women's favor in the Netherlands although this too is lacking in research.

America Individuality: As illustrated in Table 10.1 women tend to have more of a collective, team-oriented style than men. This is at odds with American individualistic traits. An example of this is how men and women differ when talking about their work contributions. Women more often use the pronoun "we" while men more often use "I." "Women are less likely than men to have learned to highlight their personal contributions. They are also more likely than men to believe that if they do so, they will not be liked" (Tannen, 1995). In American work culture if an employee is unable to demonstrate that she is working hard or that she is the one who came up with certain ideas there is a great chance that she will not score as high on her evaluations and then be considered less qualified for a promotion (Tannen, 1995).

INTERVIEWS

Six semi-structured interviews were remotely conducted. All the participants were either of Dutch or American nationality and, as selection criteria, have worked in both the United States and in the Netherlands. Interviews were aimed to gain insight from Americans on how they thought Dutch work culture was different or the same, how their own work experiences compared in the two countries, and to see if the Dutch noticed parallel differences or similarities when they worked in America. The focus was on workplace culture and, when possible, its impact on women's career advancement. To have a more complete picture of the cultural differences, some questions were designed to fill in gaps from the literature and others were to see if the literature matched the participants' experiences.

Questions were of a comparative nature and were posed for both the time worked in the United States and in the Netherlands. The interviews on workplace culture focused on the topics of directness in language, work life mentality, levels of collaboration, consensus, part-time work, maternity and paternity leave, and how success is defined in the United States versus the Netherlands.

FINDINGS

The findings showed that when it comes to the amount of time worked and level of commitment made in the workplace, all interviewees found the United States to expect or require more dedication. Interviewees worked the same or more hours in the United States than they did in the Netherlands. They were expected to work on some weekends, usually ate lunch at their desk while working, and responded to emails when taking vacation days. Interviewee 4 remarked "It is standard that I work every weekend, and in the policies, it says that we are entitled to vacation time that, if approved, means time off from work, but in reality, even that is not safe. If there is a project deadline it does not matter what else is going on. I was even expected to work on Thanksgiving [a United States national holiday] because there was a deadline." In the Netherlands, on the other hand, most participants worked a standard 40-hour workweek were never expected to work on the weekend and only had to answer emails on vacation when it was considered an emergency.

Similar to trends regarding time and commitment, participants on average received over double the vacation days in the Netherlands than they did when they worked in the United States. However, contrary to the literature, each participant took all their vacation days whether they were working in the United States or in the Netherlands. Although, Interviewee 3 remarked that "I'm still not really used to taking my leave. I guess it is the American in me. But my company here [in the Netherlands] issues forced leave you know at the end of the year. Something crazy like forcing people to take like three weeks off at Christmas just so you don't lose it [lose annual leave]." It was additionally noted by Interviewee 5 that "In America if you do not take all of your vacation time some of it will roll over to the next year. I don't remember how much, but that it is capped and anything beyond that you just loose and it's no big deal."

A strong commitment to work was again reinforced by participants when asked to define success, echoing the American "work is life" mentality previously highlighted. Every participant mentioned that both title and money are factors in the United States that make them feel successful. Other descriptors identifying success were similarly outward focused, such as owning expensive items like a "phone, car, computer, or owning other luxury material items." When the participants defined what was considered being successful in the Netherlands, only Interviewee 2 mentioned job title. Others mentioned "happiness and fulfillment, ability to enjoy life, having a happy balanced life, and, if of the right age, having a 'traditional family.'" However, when asked if work or job title help shape an individual's persona or social status in the Netherlands, responses were mixed. All participants, however, agreed that work and job title define a person's persona and social status in the United States.

Findings were less decisive regarding levels of collaboration and individual drive for both the United States and the Netherlands. This could be the result of industry or company structure and incentives. Interviewee 4 remarked that in the United States at their company "We have a strict promotion structure here. Which is based on a certain number of years at each level. That is strictly followed, yeah, I have not heard of anyone moving up faster, which I guess makes people not feel like they have to suck up extra or show off." The need for the individual mentality is less pronounced since that will not change the rate of promotion. Rather, a focus on the collective mindset of "I win when everyone wins" ensures yearly bonuses are in reach when the company is doing well. Another remarked that the Netherlands has a stronger mentality to leave the office at the set time even if there was a big deadline or to push back on tasks that seemed to be out of the scope of their job description. Interviewee 3 mentioned "I'd be on a group call right before a deadline and be talking about what needed to be done and I'd have [some] team members push back and say things like 'that task isn't my job'." [Interviewer:] "Who would do that work? Or would it not get done?" [Interviewee 3:] "Usually me, or whoever I could get to stay." Contrastingly, when Interviewee 3 worked in the United States, they felt it was always "all hands on deck" in a comparable situation. Everyone involved was expected, but also willing, to stay and help to make the deadline regardless of the time, or if the duties asked were outside of their job description. While it was also noted that Americans were willing to help the team or stay longer, it was proposed that this was mostly to show they were dedicated or to be attached to a certain project for later acknowledgment and gain.

While views on individuality versus group mentalities ultimately differed, comparing directness between the United States and the Netherlands yielded more agreement. All participants stated that the Dutch were direct. Americans had two distinct findings that varied between being politically correct or a mix of direct and being politically correct depending on the boss. Every participant agreed that Dutch women were more direct in their language than American women and that Dutch women were also equally as direct as Dutch men. It was inconclusive, however, whether Dutch men and American men are equally as direct as each other. An even split of participants felt American men were more direct, Dutch men were more direct, or that they were equally as direct. While American women were found to be less direct, there was consensus that American women became more direct as they progressed in their careers to higher positions.

Aligning with the literature, all the interviewees stated that, in most cases, a consensus was required when making a decision in the Netherlands. The only exception that was mentioned by Interviewee 6 was "I have a bit of autonomy now. But that comes with a good relationship with your manager and just being there long enough that they trust you and know how you work." This stemmed more from trust with his immediate supervisor and having worked in the company long enough to understand what would be in line with past decisions, but also have a good understanding of which decisions would need to be discussed by the whole team. All participants noted that it is acceptable in the Netherlands at any level for an employee to question a decision. However, when presenting an alternate view, there was an etiquette described by participants. Interviewee 1 stated that "[An] employee needs to provide hard facts in the favor of his/her idea and if this makes sense, the decision will be changed. New ideas are definitely allowed and valued." This tendency was also noted in the context of directness. In the Netherlands, meetings are more structured and most ideas, even when brainstorming, are presented by facts rather than business intuition.

Participants were more mixed in their views regarding decision-making in the United States. Those interviewed noted that it depended on the manager, the urgency, and size of the decision. Several participants reflected that when a quick decision was needed, a manager was often expected to step in and make a fast, clear, and final decision on what to do. However, when a decision was complex and had lots of steps or intricate parts, then it is common to have the whole teamwork out the best approach and agree to move forward.

The findings for part-time work were very much in line with what the literature stated. All the interviewees had the option to work part-time in the Netherlands, but it was either not an option in the United States or it was very rare. It was noted by Interviewee 5, "I actually feel that you could work part-time [in the United States] if you *needed* to." Working part-time is an option to most employees that are in good standing with the company but only in extenuating circumstances and usually only on a temporary basis. For example, after a medical emergency and after sick leave has been used up, part-time work may be an option in the United States. While there are laws, particularly in the Netherlands, that protect part-time workers, all participants felt that an employee who is working part-time does not have the same career advancement opportunities. This was consistent for both the Netherlands and the United States. All participants felt that the employee would have to return to working full-time if the aim was career advancement. However, multiple individuals mentioned that working part-time in the United States could have negative consequences to career advancement, even after returning to full-time work. None of the participants noted that there was any distinction here with gender and felt that male or female would have to switch back to full-time work if they wanted to advance their career.

When the topic of maternity leave came up the findings were less clear. Only a few participants know the amounts of maternity leave or paternity leave that either the United States or the Netherlands company offered. However, it is noteworthy that there were mixed findings for both the United States and the Netherlands when participants were asked whether taking maternity leave would affect career advancement upon return. This is particularly interesting because there are laws in place that prohibit the discrimination of this in both the Netherlands and the United States. For both the United States and the Netherlands some participants clearly stated they thought it would negatively affect career advancement, while others stated there would be a delay to progress. Some participants felt that it would not have any affect. Most interviewees felt that in either country there was no effect on a women's career potential if she was of an age where she might decide to have children but did not have kids yet. Interviewee 2 mentioned that she felt there was significant social pressure for women in the Netherlands to work part-time after taking maternity leave:

"The expectation almost becomes that if you are a female and you have kids that you will not work full time. I would say in certain contexts there is a level of judgment against women who decide to continue working full time." There is not only an expectation from family and peers but also colleagues and managers. It was even mentioned that women who work full-time are regarded quite negatively.

Interviewees had even less knowledge of what paternity leave was offered if any. This could be an indication that paternity leave in both countries is still less common in practice. However, it should be noted that only one of the interviewees had a child. He did remark on feeling pressure as a father not to take paternity leave, being a Dutch citizen working in the Netherlands at the time. While it was offered, it was not respected among peers or managers. Yet, working full-time but with a flexible schedule was considered acceptable and even commonplace in his office. "Yes, [we have paternity leave and are even] encouraged to take it. They want to show, on the face, that they support it, but it will hurt, it will hurt you career-wise, yes."

The interviews offered more insight into the different workplace cultures, but also for the most part supported the current literature that is available. Given the small sample size, more interviews would need to be conducted to obtain a more accurate understanding of these cultural differences. Despite these shortcomings, responses from the participants did provide a glimpse of what the work environments are like in both the Netherlands and the United States, with the different cultural lenses applied by both Dutch and American participants. Findings from the interviews complemented the literature.

Discussion

Most of the interviews supported what the literature stated on the seven major differences in work culture that were identified; directness in language, work life mentality, individual or collective mentality, consensus, part-time work, maternity and paternity leave, and how success is defined.

The most notable difference, and an unexpected finding, was how direct Dutch women were perceived by the interviewees. The literature groups all women and states that they are less direct than their male peers or supervisors. However, the interviewees agreed that there were differences between Dutch women and American women and that Dutch women were much more direct and even equally as direct as their male counterparts. According to Table 10.1, directness is the only trait women do not exhibit that would allow her to work well in a Dutch work environment. Consensus in the literature is that women in the workforce are generally more people focused and placed value on a cooperative and collective environment. This could mean that direct American women and Dutch women, if all other stereotypical traits exist, would be very successful at fitting into the Dutch work culture of consensus and collaborative mentality. They could potentially have an easier time than their male colleagues that do not exhibit these traits. If Dutch women are in fact as direct as Dutch men, then this perceived limitation as presented by literature may not be the case in the Netherlands. This could be a bigger problem for a typical American woman, who is less direct, when trying to work in the Netherlands.

However, both maternity leave and part-time work, though to a lesser extent, hinge on whether a woman wants to have children. This is a personal choice the family makes. The decision has implications for the women's career and which country is a better fit. The findings are not on whether having children affects a women's career but rather based on how different family lifestyle choice fits into which work climate better.

Another consideration is how success is defined and work life mentality. In the United States, the mentality is that "work is life" and you are considered successful when you have a good job title and make enough money to buy luxury items. This can be considered a lifestyle preference. Someone might prefer the Netherlands mentality of work life balance and the aim to be happy and fulfilled through other means besides work, but not everyone will. There are studies that indicate that the mentality in the Netherlands is "healthier" (Haar et al., 2014). This literature claims that having a better work life balance allows people to be happier and more content with their life. It also lowers levels of anxiety and depression. A healthy work life balance means not working an abundance of hours and to have the ability to disconnect when outside of working hours. As our interviewees and the literature noted, this is a particular struggle in the United States where employees are often expected to respond to emails or even continue working beyond the standard 40-hour work week. Blurring the line between work and personal time by not allowing a work life balance is the most extreme when employees are expected to work on holidays, vacations, or feel that they cannot take leave even if it is available.

The Netherlands, while not perfect on this matter, does have a better work life balance than the United States, placing more value on personal time, according to both the literature and interviewees. Working on the weekend and working long hours is less common, although in some industries weekend or long hours cannot be avoided such as retail or restaurants that are open nights and weekends. This value of work life balance also factors into the importance of family time particularly for mothers as we see with the pressure for women to spend a significant amount of time child-rearing.

It should also be considered that social constructs and ideas someone is brought up with can be hard to change. An American that moves to the Netherlands might have a harder time adjusting to the work culture versus moving from one state to another within the United States. Similarly, a Dutch person who moves and starts working in the United States might have a harder time adjusting to the work culture than if that individual moved within the Netherlands. If a Dutch employee starts working in the United States but works only the required 40 hours and does not check emails on the weekend or on vacation, this could negatively affect their career and they could be perceived as less committed than their peers that do exhibit those behaviors. An American who starts working in the Netherlands but has materialistic values could be perceived as extravagant or even greedy and money hungry, which might not fit in well with Dutch work culture. Management might not want employees in the company to hold traits that do not align with the goal and image of the company. It is not enough to identify which culture is better for women's career advancement, but rather, each nation's cultures, ideas, and mentalities will need to be embraced to achieve success in either country.

Propositions

Drawing from the literature and the interviews on various cultural and legal aspects discussed in this chapter, four propositions have been developed to answer the question "in which country is it easier for women to advance her career if she chooses to?"

Proposition 1: If a woman decides against motherhood the Netherland offers a more favorable work culture for her career advancement.

The work environment in the Netherlands is more aligned to stereo-typically female working styles. The focus on people and relationships rather than results and the collective collaborative nature plays to women's strengths. There is however the requirement that the women will need to be as direct as her male peers. Given directness is standard in Dutch work culture, a woman's overly indirect behavior could be seen as a weakness.

Proposition 2: If a woman wants children and a reduced/balanced work and home life, the Netherlands is a better place for her continued career.

Due to the guaranteed maternity leave and the availability of part-time work in the Netherlands, women legally have more rights and protec-tions. This allows the future mother to continue to work, but with the flexibility to prioritize her family as well, without making drastic sacrifices to either her career or her family. In contrast, mothers in America that want to prioritize their families often must stop working completely. It can then be difficult to reenter the workforce after a significant break (Krueger et al., 2014). Generally, only wealthy families can afford to consider reductions in the United States, because part-time work typi-cally doesn't provide enough resources to sustain a family. The family would have to be able to survive off only the father's income, meaning he would have to be making enough to afford for the mother not to work or work part-time. Whereas in the Netherlands working part-time is sufficient, especially when coupled with a partner's full-time income. This greatly contributes to the number of both men and women in the Netherlands who choose to work part-time as they can afford to do so.

Proposition 3: If a woman wants children but also wants to continue to work full time, then the United States is a better place for her career advancement.

In the United States, there is little to no societal pressure for mothers to stay at home or work part-time after starting a family. If a woman is willing to work full-time, the norm in the United States tends to be more supportive than that of the Netherlands, if she demonstrates that she is committed. As stated before, part-time work is rare and not lucra-tive enough; it is not a common choice new mothers make in the United States. Whereas in the Netherlands, because of the social pressure on mothers and the ease and availability of part-time work, the norm is that 75% of women do not continue to work full-time after having children.

> Proposition 4: In general, it appears that the Netherlands is a more balanced place to work regardless of gender due to a healthier work life balance and the way success is defined.

The goal in the Netherlands is to be happy and have a balanced life, rather than the American goal of status and materialistic success. Having this healthier balance reduces the risk of burnout and employees are better able to manage stress which can lead to an overall longer and happier career (Haar et al., 2014). It is also shown that employees can be more productive if there is a better work life balance (Haar et al., 2014). After about 50 hours a week productivity significantly drops off and working any more than 55 hours a week shows no additional outcomes, implying working 80 hours or 55 hours a week will equal the same output (Pencavel, 2014). This again supports the notion that having better work life balance would cause an employee to be less stressed, reduce burnout, reduce turnover and increase productivity which could mean better work performance overall. If burnout is reduced and work performance is better, these two things will aid in the chances of a longer sustained, productive career.

It should be mentioned, however, that there are several limitations to these propositions. For example, research showed that mothers in the Netherlands feel some pressure to limit working hours, but the effects of other societal views on gender roles and responsibilities are not considered.

In both the United States and the Netherlands, different industries and job types have more or less women than other industries. Choosing a more conventional female career could have different implications on career advancement than if a woman chose a notoriously male-dominated industry. This is true in both the Netherlands and the United States but could vary in degree. Industry specifics or differences in gender within industries between the United States and the Netherlands were not considered in this chapter. Similarly, different trends in education and part-time work by industry will also vary between the two nations.

When talking about the cultural differences, one thing that was not considered is each country's acceptance and openness to other cultures in the workplace. However, additional difficulties could be encountered for Americans working in the Netherlands and vice versa depending on a companies' ability and acceptance of differing work cultures. The discoveries made within this research, and accompanying propositions, do not factor acceptance of cultural differences in the workplace.

The biggest limitation to the interviews was the small sample size. Additionally, while participants that have worked in both the United States and the Netherlands and can compare the two, they most likely have a more international mindset, are likely more willing to travel and live abroad than the average employee. Commonly, this also means working in an international company and in a more international industry. These types of individuals do not adequately represent the majority of either the United States or the Netherlands, thus skewing the sample.

It will also be important to see if the developments of COVID-19 and the subsequent quarantines, restrictions, and work from home policies affect any of the seven cultural norms described in this chapter.

CONCLUSION

Based on the literature on workplace culture in the Netherlands and the United States, how different laws or workplace trends affect women's opportunities, and the interviews presented in this chapter, it appears that the Netherlands, in most respects, will be a more favorable place for a woman to grow and advance her career.

There are two major conditions to consider. First, from the expert interviews it emerged that directness seems to be a defining characteristic to being Dutch. However, any society will have people who are more reserved and even shy, and others that are more outgoing and forward. Someone who is Dutch but is naturally shyer and more uncomfortable with confrontation would be at odds with this norm just like someone from another culture that is not known for their directness. Women from other cultures as exampled in the United States are less direct, clear, and forward than males tend to be. This can have a negative effect on those individuals' career as they might be perceived as less qualified to lead and manage others, which could disproportionately disadvantage women.

The second consideration is the change in dynamics and expectation after a woman has a child. Interviewees report that the work environment is stable and fairly level for men and women until a woman decides to start a family. Prior to having children, Dutch women have little cultural norms or perceptions that hold them back compared to the United States.

However, for most women, equality will be reached when women are treated the same and offered the same as their male counterparts when in an equal position doing equivalent work regardless of whether she decided to have a family.

While both countries have strengths, weaknesses, and general cultural differences, this chapter looked closer at workplace environments. To truly gain a holistic perspective of how each country compares in women's career advancement, all five of the World Economic Forum's Global Gender Gap Report categories—economic participation and opportunity, educational attainment, health and survival, political empowerment—need to be considered and biases and perceptions from all angles need to be factored.

REFERENCES

Allen, R. (n.g.). *Generational differences chart.* Shepherd, Michigan: Presentation given for Staff Professional Development Day at West Midland Family Center. http://www.wmfc.org/uploads/GenerationalDifferencesChart.pdf

Barone, A. (2020, March 27). *American dream.* Investopidea. https://www.inv estopedia.com/terms/a/american-dream.asp

Belephant, T. (2017). Communication styles: Increasing awareness. *All capstone projects,* 333. Governors State University. https://opus.govst.edu/capsto nes/333

Benschop, Y., Brink, M., Doorewaard, H., & Leenders, J. (2013). Discourses of ambition, gender and part-time work. *Human Relations,* 66(5), 699–723.

Big Red Car. (2013, January 28). *The design of compensation packages for C level execs and senior management.* The Musings of the Big Red Car. https://themusingsofthebigredcar.com/the-design-of-compensation-pac kages-for-c-level-execs-and-senior-management/

Breeschoten, L., & Evertsson, M. (2019, March 1). When does part-time work relate to less worklife. *Community, Work & Family,* 606–628. https://doi. org/10.1080/13668803.2019.1581138

C., V. (2017, June 29). *Gender communication differences and strategies.* Professional Development. https://experience.com/advice/professional-dev elopment/gender-communication-differences-and-strategies/

Cassard, A. M., & Hamel, J. C. (2008). Women and their relationship to leader, follower commitment, and job performance: Netherlands, Belgium, and North Carolina. *Journal of Social, Behavioral, and Health Sciences, 2,* 68–82.

Chullen, C. L., Adeyemi-Bello, T., & Vermeulen, E. (2017). A comparative analysis of attitudes towards women as managers in the U.S. and Netherlands. *Journal of Leadership, Accountability and Ethics, 14(2),* 24–42.

Chung, H.-J., & Lippe, T. V. (2018). Flexible working, work–life balance, and gender equality: Introduction. *Social Indicators Research,* 1–17. https://link.springer.com/article/10.1007/s11205-018-2025-x

Clockify. (2020). *Standard working hours (2020).* Clockify working hours. https://clockify.me/working-hours

Conway, L., Repke, M., & Houck, S. (2017, May 10). Donald Trump as a cultural revolt against perceived communication restriction: Priming political correctness norms causes more Trump support. *Journal of Social and Political Psychology, 5(1),* 244–259.

Crotti, R., Geiger, T., Ratcheva, V., & Zahidi, S. (2020). *Global gender gap report 2020.* World Economic Forum.

Dennison, J. (n.d.). *Why new dads should take paternity leave.* Parents. https://www.parents.com/pregnancy/my-life/maternity-paternity-leave/why-new-dads-should-take-paternity-leave/

Dikkers, J., Engen, M., & Vinkenburg, C. (2010, October 26). Flexible work: Ambitious parents recipe for career success in the Netherlands. *Career Development International, 15(6),* 562–582.

ET Online. (2018, June 19). 5 ways to ensure your employer sees you as 'go-getter' and not a 'lagger'. *The Economic Times.* https://economictimes.indiatimes.com/jobs/5-ways-to-ensure-your-employer-sees-you-as-go-getter-and-not-a-lagger/articleshow/64646345.cms?from=mdr

Evanson, N. (2016). *Dutch culture.* Cultural Atlas. https://culturalatlas.sbs.com.au/dutch-culture/dutch-culture-communication

Expat insider 2018. (2018). https://www.internations.org/expat-insider/2018/

Gallup, Inc.; Populace, Inc. (2019). *Success index.*

Haar, J. M., Russo, M., Suñe, A., & Ollier-Malaterre, A. (2014). Outcomes of work—Life balance on job satisfaction, life satisfaction and mental health: A study across seven cultures. *Journal of Vocational Behavior, 85(3),* 361–373. https://doi.org/10.1016/j.jvb.2014.08.010

Hess, A. (2018, July 6). *Here's how many paid vacation days the typical American worker gets.* CNBC. https://www.cnbc.com/2018/07/05/heres-how-many-paid-vacation-days-the-typical-american-worker-gets-.html#:~:text=In%202017%2C%20the%20average%20worker,given%2020%20paid%20vacation%20days

Hofstede. (2010). *Country Comparison.* Hofstede Insights. https://www.hofstede-insights.com/country-comparison/the-netherlands,the-usa/

Hofstede, G., Hofstede, G. J., & Minkov, M. (2010). *Cultures and organizations: Software of the mind: Intercultural cooperation and its importance for survival* (3rd ed.). McGraw-Hill.

International business guides: Netherlands. (2020, January). HSBC. https://
www.business.hsbc.com/business-guides/netherlands?cid=HBEU:AH:NU:
P0:CMB:L14:XXG:XTR:14:XX:13:0323:016:IBG2020&ds_rl=1280576&
gclid=CjwKCAjw0_T4BRBlEiwAwoEiAXWpmeStyyynrFoChiwNfzYrNaK4
w6nlQ4kVBSWU5WrvBg_RXuu4JBoCRVYQAvD_BwE

International business guides: The United States. (2020, January). HSBC.
https://www.business.hsbc.com/business-guides/us?cid=HBEU:AH:NU:P0:
CMB:L14:XXG:XTR:14:XX:13:0323:020:IBG2020&ds_rl=1280576&gclid=
CjwKCAjw0_T4BRBlEiwAwoEiAZGExhVvccrY9Gi09CJgTOIgCUhAxptv7n
kEgn1LR-q0xF1bB2ME3RoCoRIQAvD_BwE

Johns, M. L. (2013). Breaking the glass ceiling: Structural, cultural, and organizational barriers preventing women from achieving senior and executive
positions. *Perspectives in Health Information Management, 10* (winter, 1).

Jong, K. (2018, December 5). *Work culture in the Netherlands.* Career Professor.
https://careerprofessor.works/work-culture-in-the-netherlands/

Kemp, R. V. (2018). *Dutch work culture.* The Hague International Centre.
https://www.thehagueinternationalcentre.nl/relocating/work-in-the-nether
lands/dutch-work-culture

Kim, A. (2019, August 19). *A record 768 million US vacation days went to waste
last year, a study says.* CNN. https://edition.cnn.com/travel/article/unused-
vacation-days-trnd/index.html

Krueger, A. B., Cramer, J., & Cho, D. (2014). Are the long-term unemployed
on the margins of the labor market? *Brookings Papers on Economic Activity,
Spring*(1), 229–299. https://doi.org/10.1353/eca.2014.0004

Lalor, A. (2020, January 10). *Becoming a father? Here's everything you need
to know about paternity leave in the Netherlands in 2020.* Dutch Review.
https://dutchreview.com/culture/becoming-a-father-heres-everything-you-
need-to-know-about-paternity-leave-in-the-netherlands-in-2020/

Mobility, E. I. (2020, July 13). *Understanding Dutch business culture.* Expatica.
https://www.expatica.com/nl/working/employment-basics/dutch-business-
culture-102490/

Netherlands Enterprise Agency, RVO. (2020). *Leave schemes.* https://business.
gov.nl/regulation/leave-schemes/

Paid Vacation Days Europe. (2021, January 22). Euro Dev. https://blog.eur
odev.com/paid-vacation-days-europe-2021

Parmelee, M. (2020). *The Deloitte global millennial survey 2020: Millennials and
Gen Zs hold the key to creating a "better normal".* Deloitte.

Pencavel, J. (2014, June 3). The productivity of working hours. *The Economic
Journal, 125*(589).

Robert Half. (2020, March 0). *The value of teamwork in the workplace.*
Robert Half Blog. https://www.roberthalf.com/blog/management-tips/the-
value-of-teamwork-in-the-workplace

Robertson, M., & Vink, P. (2012). Examining new ways of office work between the Netherlands and the USA. *Work: A Journal of Prevention, Assessment & Rehabilitation, 41*(Supplement 1), 5086–5090.

Rodriguez, A. (2017, December 26). *The American Dream is not set up for the poor to succeed.* Medium. https://medium.com/@PStoplight/the-american-dream-is-not-set-up-for-the-poor-to-succeed-cfa4da39f16e

Ross, M., & Bateman, N. (2019, November 21). Low-wage work is more pervasive than you think, and there aren't enough "good jobs" to go around. *The Avenue.*

Selvarajah, C., Meyer, D., Waal, A. A., & Heijden, B. I. (2012). Characteristics of high performing managers in The Netherlands. *Leadership & Organization Development Journal, 33*(2), 131–148.

Steck, A. (2020, January 16). *Everything you need to know about maternity leave in The Netherlands in 2020.* Dutch Review. https://dutchreview.com/expat/health/maternity-leave-in-the-netherlands-2020/

Tannen, D. (1995). The power of talk: Who gets heard and why. *Harvard Business Review, 73*(5), 138–143.

The Netherlands: Dutch business culture. (2019). Business Culture. https://businessculture.org/western-europe/business-culture-in-netherlands/

Thomas, M. (2016, September 6). The Impact of Mandated Maternity Benefits on the Gender. Retrieved from http://digitalcommons.ilr.cornell.edu/ics/16

Tidey, A. (2020, February 19). *Brexit: Number of companies relocating to the Netherlands 'is accelerating'.* Euro News. https://www.euronews.com/2020/02/19/brexit-number-of-companies-relocating-to-the-netherlands-is-accelerating

Turner, C. (2012, May 7). A balance of both masculine and feminine strengths: The bottom-line benefit. *Forbes.*

U.S. Bureau of Labor Statistics. (2018, June 28). *TED: The Economics Daily image.* Bureau of Labor Statistics. https://www.bls.gov/opub/ted/2018/private-industry-workers-received-average-of-15-paid-vacation-days-after-5-years-of-service-in-2017.htm

U.S. Bureau of Labor Statistics. (2020, July 28). *Labor Force statistics from the current population survey.* U.S. Bureau of Labor Statistics. https://www.bls.gov/cps/lfcharacteristics.htm#fullpart

U.S. Department of Labor. (n.d.). *Family and Medical Leave (FMLA).* Department of Labor. https://www.dol.gov/general/topic/benefits-leave/fmla

U.S. Department of Labor Wage and Hour Division. (2020). *Overtime pay.* Department of Labor. https://www.dol.gov/agencies/whd/overtime

Wielers, R., & Raven, D. (2013). Part-time work and work norms in the Netherlands. *European Sociological Review, 29*(1), 105–113.

Work-life balance. (2019). Business Culture. https://businessculture.org/western-europe/business-culture-in-netherlands/work-life-balance-in-netherlands/

Zintl, A. (n.d.). *Know your FMLA maternity leave rights.* Parents. https://www.parents.com/pregnancy/my-life/maternity-paternity-leave/maternity-leave-rights/

Respect or Transgression of Norms in the Context of Religious Diversity: The Example of the Mormons in France

Victoria Bacouel and Sabine Jentjens

Abstract Public controversies can be understood as a confrontation between various systems of norms. Rather than confirming a given moral order, these types of controversies may provoke a moral positioning and help in clarifying lines of difference or conflict in a society. Based on a qualitative study, this chapter examines how the controversy over the construction of the Mormon temple in France impacted the perception of Mormons as a potential religious out-group in France. We are interested whether French Mormons are seen to respect both legal and social norms in France or whether they are perceived as transgressors of one or

V. Bacouel
Rotterdam School of Management, Rotterdam, The Netherlands
e-mail: vic.bacouel@laposte.net

S. Jentjens (✉)
ISC Paris Business School, Paris, France
e-mail: sabine.bacouel-jentjens@iscparis.com

A. S. Arora et al. (eds.), *Managing Social Robotics and Socio-cultural Business Norms*, International Marketing and Management Research, https://doi.org/10.1007/978-3-031-04867-8_11

the other. In doing so we contribute to the broader discussion on religious diversity and anti-discrimination in France, a country that declares itself to be secular.

Keywords Socio-cultural norms · Legal norms · Transgression · Mormons · Religious diversity · France

INTRODUCTION

Fans of fantasy literature probably first heard about Mormonism during the worldwide success of the Twilight saga, written by Mormon Stephenie Meyer. The author's religion has led the international press to analyze the content of Twilight for conscious or unconscious propaganda of Mormonism principles (Aleiss, 2010; Atlantico, 2012; Heath, 2011). Those who are not interested in literature, but in international politics, may have heard about Mormonism at the latest during the 2012 US presidential election. For the Republican and Mormon candidate, Mitt Romney, was facing an electorate divided over whether Romney's religion, was of Christian faith (Smith, 2016). Frequent travelers may have already found in a drawer of their room at the Marriott hotel, next to the Bible, the Book of Mormon, as the American founder of this hotel chain is also a Mormon.

If the examples above refer exclusively to the United States, the cradle of Mormonism, one should not forget that the number of Mormon followers is rapidly growing worldwide, with about 36,0000 followers in France (Eglise de Jésus-Christ des Saints des Derniers Jours, 2011).

However, interest in Mormons in France has only recently gained public attention. French Mormons did not become a topic of debate until 2012 with the construction and opening (in 2017) of the first Mormon temple in mainland France, located in Le Chesnay, near Versailles. While the construction of the Mormon temple sparked a public controversial debate, a Mormon also influenced the 2017 French presidential campaign. Indeed, the opponents of the Republican candidate, François Fillon, reproached him for having hired as a tax advisor "the Mormon" Dominique Calmels, otherwise, in charge of Mormon communication in France and a volunteer priest within the Mormon community in Paris (Goosz, 2017).

In the context of these sudden public controversies over this religious group, it is interesting to investigate the image of Mormons in France who form a part of the French society and participate in the occupational and public life and contribute to the religious diversity of the French society. France declares itself as a secular society where one's religion remains in the private sphere (Bertossi, 2011). Recent discussions on religious diversity in France were focused on Islam which arouses concern and general fear, especially since the latest terrorism attacks in France (Yang & Bacouel-Jentjens, 2019).

Public controversies can be understood as a confrontation between various systems of norms. Rather than confirming a given moral order, these types of controversies may provoke a moral positioning and help in clarifying lines of difference or conflict in a society (Jacobsson & Löfmarck, 2008). Based on a qualitative study, we examine how the controversy over the construction of the Mormon temple in France impacted the perception of Mormons in France. We are interested whether French Mormons are seen to respect both legal and social norms in France or whether they are perceived as transgressors of one or the other and if this leads to religion-based discrimination.

First, we introduce our theoretical concept of social norms, deviance, and transgression. We present this concept in a religious context to know in which form a membership to a denomination is defined as deviant with respect to social and/or legal norms. Then, we focus on the Mormon denomination in general and in France and we expose the history and the main ideas of Mormonism. In a second step, we carry out an empirical work on the perception of Mormons in France based on interviews with eight Mormon and non-Mormon informants in France.

THEORETICAL BACKGROUND: FRENCH MORMONS IN FRENCH SOCIETY

Norms and Deviance

The norms of a society are rules of conduct specific to that society. They are legitimized by the values of that society and learned and shared by its members. A distinction can be made between social norms and legal norms (Passard & Perl, 2013). Social norms are learned by the individual through socialization. They do not come from nature or from a universal morality but from a social construct. Therefore, they are variable across cultures and over time. A legal norm is everything that is of the order of the law. These norms may or may not overlap. For example, greeting your neighbor is a social norm that is not sanctioned by law. Monogamy in France is a social norm and at the same time a legal norm because polygamy is sanctioned by the law. Finally, the law prohibits speeding in France, but the social norm tolerates it (Passard & Perl, 2013). The domain of social norms is much broader than that of legal norms.

We speak of deviance if the social and legal norms in force in a society are transgressed and if this transgression involves sanctions. Deviance is a two-step process. The first stage is the transgression of the norm. This is called primary deviance. The second stage, secondary deviance, corresponds to the recognition of a deviant act by an authority of social control and this leads to a stigmatization of the deviant individual or group (Mucchielli, 1999). According to the American sociologist, Howard S. Becker, deviance is a construction of society. These are groups of individuals who defend social norms even in the absence of an institutional demand (such as the law) (Becker, 1963). As deviance is socially generated it becomes inevitable in any society. Even in a society of saints there must be sinners to affirm the rules among those untouched by transgression and rekindle the solidarity of the society. In other words, the social structure fosters transgressions to periodically foster the collective outrage that would affirm the sanctity of norms. At the same time, transgression can be a basis for social change as certain laws and norms are contested. At periods of social change marked by anomie, deviance may become the basis for innovation and indeed, adaptation to new moral circumstances. What is considered transgressive in one era often becomes acceptable in another era and normative even later (Durkheim, 1982).

Religious Values—Norm-Matching or Deviant?

Religion is a system of beliefs and values that provides followers with answers to the meaning of life, death, and suffering. Religion is composed of elements that allow the believer to externalize a religious feeling in a communal and formal structure. In the various religions, there is a system of values in harmony with the respective beliefs, shared by the whole religious community and maintained by moral obligations and ethical rules. Rituals, ceremonies, and rites of passage have a cultural component (Derocher, 2018). Traditional religions in the world include Christianity, Islam, Hinduism, Judaism, and Buddhism (Laffargue, 2013).

Traditional religions have been around for a long time and are strongly attached to the culture of a people. New religious movements offer, like traditional religions, a community, a place to express spirituality, and a system of values and norms. However, a new religious movement rarely identifies with a culture. Often their beliefs are at odds with the doctrines of traditional religions. Some new religious movements prescribe little to no norms, while others establish a strict and rigid set of rules (Derocher, 2018). However, this term of new religious movements is controversial. The term "new" to refer to certain minority religious movements such as Jehovah's Witnesses or Mormons is misleading, as these movements have been around since the nineteenth century. On the other hand, the term "religious" is rejected by some movements that define their practices as spiritual, not religious. Conversely, critics of Scientology deny it any religious character (Mayer, 1989).

Cults also offer a community, a belief system, and a framework for living spirituality within a system of values and norms. However, the cult finds itself in conflict with the values of the society in which it lives by rejecting the dominant values and withdrawing into itself, causing a radical break with society. Cults challenge, to varying degrees, certain components of the society around them such as the political, legal, medical, educational, or scientific systems as well as values of equality, individualism, freedom, and economy (Derocher, 2018). Some sectarian groups choose physical isolation to proscribe contact with non-members. They thus constitute micro-societies (Derocher, 2018).

French law does not provide a legal definition of either a cult or a religion (Miviludes, 2018) in relation to the principle of secularism and in order not to offend the freedom of conscience, opinion, and religion guaranteed by the Declaration of the Rights of Man and of the Citizen of 1789, the French Constitution of October 4, 1958, and the 1905 law separating the Churches and the State. Nevertheless, the law sets limits that sanction the abuse of these freedoms. This approach of the French legal system can, therefore, be summarized by reference to Article 10 of the Declaration of the Rights of Man and of the Citizen of 1789, which states that "No one shall be disturbed on account of his opinions, even of a religious nature, provided that their manifestation does not disturb the public order established by the law."

The French legal system is, therefore, not concerned with cults and beliefs as such. It is intended to be both pragmatic and textually framed: it aims at the prevention and repression, not of cults per se, but of sectarian aberrations. To observe and analyze the phenomenon of sectarian aberrations, to inform the public about the risks they represent, and to coordinate the preventive and repressive action of the public authorities, the Inter-ministerial Mission of Vigilance and Combat against Sectarian Aberrations (MIVILUDES) has been instituted as an inter-ministerial mission under the Prime Minister (MIVILUDES, 2018). MIVILUDES receives approximately 2000 reports of aberrations per year. Based on its experience and on established jurisprudence, both in the sphere of private law and administrative law, this organization defines a sectarian aberration as follows:

> It is a deviation from the freedom of thought, opinion, or religion which undermines public order, laws or regulations, fundamental rights, security or the integrity of individuals. It is characterized by the implementation, by an organized group or by an isolated individual, whatever its nature or activity, of pressure or techniques aimed at creating, maintaining or exploiting in a person a state of psychological or physical subjection, depriving him or her of part of his or her free will, with harmful consequences for that person, his or her entourage or for society. (MIVILUDES, 2018: https://www.derives-sectes.gouv.fr/quest-ce-quune-d%C3%A9rive-sectaire)

Freedom of religion is a fundamental element of human rights in international law. Article 18 of the United Nations Universal Declaration of Human Rights states that every human being has the right to freedom of thought, conscience, and religion; this right includes freedom to change one's religion or belief, and to manifest his religion or belief, either alone or in community with others and in public or private, in teaching, practice, worship, and observance. In Europe, freedom of religion is guaranteed by Article 9 of the European Convention on Human Rights (ECHR), which applies to all member states of the Council of Europe. In case of violation of the ECHR by a signatory state, the European Court of Human Rights can be called in Strasbourg.

One could, therefore, conclude that the religious movements presented above (traditional religions, new movements, and cults) do not transgress the legal norms of countries that respect the United Nations Universal Declaration of Human Rights. On the other hand, according to a study by the American research center The Pew Center, freedom of religion is restricted or even denied by law in 157 countries (Rathgeber,

2013). In these countries, a religious movement other than that permitted by law transgresses the legal norm.

Also, we can conclude that traditional religions as well as new religious movements do not transgress social norms if they respect the values of the respective culture of the country. According to the above definition, the cult conflicts with the values of the society in which it lives by rejecting the dominant values. It, therefore, transgresses social norms and, in the case of an overlap of social and legal norms, also legal norms.

Religion in the French Diversity Debate

France separates the public and private spheres as well as the state and church and expression of religious differentiation must remain in the private sphere (Bertossi, 2011). In other words, France is a secular country, and religion and work are strictly separated. Individuals are expected to assimilate universal national values regardless of their religious identity (Heckmann & Schnapper, 2003). Therefore, France has refrained from collecting statistics based on religious affiliation and it is forbidden for administrations and employers to record religious affiliation. 45% of French people describe themselves as agnostic or atheist (INED-INSEE, 2010) and the French religious landscape is characterized by secularization; nevertheless, there has been the emergence of Muslim religion, mainly among immigrants and descendants of immigrants. European immigrants have mostly integrated smoothly into French society. It is suggested that customs of European Christians made identification with French culture easier (Vladescu, 2006). Asian immigrants are generally perceived as unobtrusive (Gayral-Taminh, 2009) as Confucian principles of respect facilitate their social integration (Gayral-Taminh, 2009). However, Muslim immigrants increasingly meet strong feeling as French link Islam, especially since the rise of Islamic terrorism, with fundamentalism which is seen as opposition to French values (Vladescu, 2006). However, Mormons do rarely enter the immigrant category and have not been perceived that much as a visible minority in France.

The Mormon Community in France

As mentioned above, Mormons are often classified in the literature as a new religious movement.

The Foundations of Mormonism

Mormons are the followers of the "Church of Jesus Christ of Latter-day Saints," a church founded in 1830 in the United States by Joseph Smith. The headquarters of the church is in Salt Lake City, Utah, USA. Mormon belief is based on both the Old and New Testaments, and the Book of Mormon, which is a collection of visions delivered by the angel Moroni to Joseph Smith. Followers consider faith in Jesus Christ and his atonement to be fundamental principles of their religion. Because of some doctrinal differences, Catholics, Orthodox, and many Protestant churches consider the church to be distinct and separate from traditional Christianity (Gillette, 1985).

Mormon doctrine is based on continuing revelation. Thus, the president of the Church is a prophet, seer, and revelator for the modern age. It is Jesus Christ who, under the direction of God the Father, directs the Church by revealing His will to this president. Individual members of the Church believe that they can also receive personal revelation from God in the conduct of their lives. The president leads a hierarchical structure with various levels down to the local congregations. Bishops lead local congregations. Male members may be ordained to the priesthood, provided they meet the standards of the church. Women are not ordained to the priesthood but may serve in leadership roles in some auxiliary church organizations (Rigal-Cellard, 2003). Men and women can serve as missionaries. The Church maintains an extensive missionary program that includes promoting and conducting humanitarian services around the world.

The congregation follows a health code that abstains from alcohol, coffee, tea, and tobacco and recommends a moderate consumption of meat and grains. Mormons also adhere to a moral code, called the "law of chastity" which prohibits adultery, homosexual behavior, and sexual relations outside of marriage. In 1830, it was common for a Mormon man to marry several women. In 1890, however, the Mormon Church decided to end polygamy.

Today, the Mormon church boasts 16 million members worldwide (Europe1, 2017).

The Evolution of the Mormons in France

The first Mormons settled in France in 1849. But the conversion of new disciples proved difficult. The lack of French-speaking missionaries and a supposed lack of spiritual receptivity of the French were mentioned as reasons. The first president of the French Mormons, Louis Bertrand, wrote in 1863 to Brigham Young, the successor of Joseph Smith: "[...] nothing can be hoped from the unfaithful French: they are all spiritually dead" (Rigal-Cellard, 2000: 7).

The literature reports an anti-Mormon campaign in France after World War II and a refusal by the French government to grant residence permits to foreign Mormon missionaries. Rumors say that the Mormon Church threatened Minister Robert Schuman with pressuring the US government to stop the Marshall Plan financial aid if the anti-Mormon campaign did not stop (Rigal-Cellard, 2000). In the post-war period, the Mormons began an information campaign in France with conferences, radio concerts, and intensive proselytizing using, among other things, a missionary basketball team which, after each game won, spoke about their Church to their fans (Rigal-Cellard, 2000). Professor Bernadette Rigal-Cellard, a specialist in contemporary North American religions, particularly Mormonism, at the University of Bordeaux Montaigne, notes: "If the reputation of the Church improved somewhat, it would not attract conversions: in 1951 there were only 116" (Rigal-Cellard, 2000: 8). According to official sources of Mormons in France, their number in metropolitan France today amounts to 36,000 members (Church of Jesus Christ of Latter-day Saints, 2011). Since 1952, the Church in France has been registered as an association under the 1901 law by the Ministry of the Interior.

The Mormon Temple in Le Chesnay

A Mormon church and a Mormon temple differ in the level of sacredness of these two types of places of worship. Churches are open to the public, while temples are open only to Mormons with a pass. Churches welcome congregants for Sunday worship and social activities (Europe1, 2017). In contrast, the temple is the most sacred place for Mormons and reserved for individual or family sacraments with eternal significance, such as the sealing for eternity of the couple (marriage), children, and ancestors (baptism). Mormons erect temples only when the number of worshipers in the territorial area exceeds a certain threshold. As of 2018, there are 161 temples in the world, with 10 under construction, compared to several thousand churches (Temple List, 2018). Until 2017, metropolitan France had 110 Mormon churches but no temples. French Mormons were attached to temples in Madrid or Zurich (Rigal-Cellard, 2003).

The temples are exceptionally accessible to everyone when they are under construction or have just been built, before being consecrated (Rigal-Cellard, 2003). To gain access to the temple, the worshiper must show a "recommendation card" which she/he obtains only if she/he has satisfied the requirements of his bishop and the president of his stake (a stake is an administrative division and is composed of a certain number of wards). This pass is only valid for one year and is conditionally renewable. The leaders make sure that the worshiper is active in his or her stake, pays tithes and complies with all the community standards.

As mentioned before, French Mormons did not have a temple in mainland France until 2017. They had to travel abroad for the sacraments (weddings, baptisms) which can only be done in a temple. Since 1998, French Mormons had been looking for land in the Paris region to build a temple (L'Obs, 2017). After several searches, they found a wasteland in the commune of Le Chesnay, which is only two kilometers from the Château de Versailles. After validation of the building permit and the failure of four appeals filed before the Administrative Court by disgruntled residents, construction work began in 2012. Five years later, in 2017, the temple with parking, garden and hotel residence with a total surface of 9000 m^2, and a cost of 80 million euros was inaugurated in the presence of Mormon leaders. This construction caused a lot of controversy in the French press, between the town hall of Le Chesnay and the opposing residents and political opposition in the town. This local event in Le Chesnay put Mormonism into the focus of the French press. It is, therefore, interesting to analyze how the debate on the Mormons' temple impacted the perception about Mormons' respect or deviance of social and legal norms. As mentioned earlier, society fosters transgressions to periodically foster the collective outrage that would affirm the sanctity of norms. In the underlying research context, the construction of the Mormon temple, its press coverage, and some opposition in front of this project reflected this outrage.

METHODOLOGY

A qualitative study approach was used to examine perceptions of Mormons and non-Mormons on their acceptance within the French society with regard to whether French Mormons are seen to respect both legal and social norms in France or whether they are perceived as transgressors of one or the other. Despite challenges of generalization beyond the perception of the investigated informants (Yin, 2011), this approach would be useful for the investigation of divergence perception of certain groups within a society and their potential discrimination, thanks to the in-depth contextual information provided.

Data Collections and Analysis

Data collection was based on episodic interviewing (Flick, 2000) to activate respondents to select relevant situations and to describe what makes these episodes relevant to them. This method permits the researcher to obtain insights about what the respondent sees as significant and acceptable to communicate. The method is also useful to analyze how knowledge shared by a specific group is different from knowledge shared in other groups (Flick, 2000). The evidence presented in this paper was generated through eight semi-structured interviews with informants concerned with Mormonism and the construction of the temple in Le Chesnay (Table 11.1). As part of our study, we found it appropriate to speak with some of the stakeholders involved in the construction of the Mormon temple. We wanted to determine whether any rejection of the temple construction was related to discrimination against Mormons, and

Table 11.1 Participants in the qualitative study

Informant	Fonction	Gender	Religion
Mayor	Mayor of Le Chesnay since 2014	M	Non-Mormon
French Mormon speaksperson	Speaker of the Mormons in France and Mormon priest	M	Mormon
French Mormon	Member of the French Mormon community since 1979	F	Mormon
French missionary	French Mormon missionary in France	F	Mormon
American missionary	American Mormon missionary in France	F	Mormon
President cult defense	President of the Association for the Defense of Families and Individuals Victims of Cults (ADFI des Yvelines)	F	Non-Mormon
French citizen 1	French neighbor to the Mormon temple	F	Non-Mormon
French citizen 2	French citizen of Christian religion	M	Non-Mormon

if so, what the reason was. We conducted interviews with the mayor of Le Chesnay, who granted the building permit for the temple, Mormon spokespersons in France, two Mormon missionaries in France, the president of the Association for the Defense of Families and Individuals Victims of Cults (ADFI), and two residents near to the temple.

For each interview, we developed an adapted interview guide that consisted of a list of questions about the interviewee's function, his or her role in the project to build the Mormon temple in Le Chesnay, the evolution of this perception, and finally his or her personal image of Mormons in France. For the representatives of the Mormon community interviewed, we were interested in how they felt perceived and accepted in France. The interviews lasted between thirty and ninety minutes. Six interviews were conducted in French and two in English. They were recorded and transcribed.

Analysis of interviews involved three steps: (1) Identification of episodes in the interview files (word documents) and extraction of those episodes that clearly qualified as perception on Mormons as accepted or discriminated minority in France. Episodes typically involved descriptions of experiences of the informant. (2) Thematic selection of episodes for examining on what features of norms (legal or social) the episode focused. (3) Coding of episodes to identify positive or negative perceptions of Mormons' respect of norms in and potential discrimination by the French society.

The Perception of Mormons in France in the Context of the Construction of the Mormon Temple in Le Chesnay

If one launches an internet search on Mormons in France, one quickly realizes that French Mormons attract the interest of the media only occasionally. This is particularly true in 2017, when the first Mormon temple in France was inaugurated in Le Chesnay near Versailles.

Do What You Want as Long as You Respect the Law

Our results show that the perception of Mormons is predominantly based on the compliance with legal norms. If Mormons as a cult or religious group behave in line with the French laws, they are not seen as deviant by French officials and citizens. French do not seem to be particularly informed about Mormons. However, the construction of the temple led to a temporarily focus on the Mormon community and to an involvement of those who were primarily concerned: authorities and residents.

The mayor of Le Chesnay reports that the first thing he did when he learned of the project to build a Mormon temple in his commune was to check that the legal standards were respected not only with regard to the construction but also with regard to the legal status of Mormonism in France:

> When the Mormons arrived in my office, the first thing I did was to create a commission with the elected officials to find out who they were. I had no idea who the Mormons were. (Mayor)

According to him, a value of the French republic is secularism and his duty as mayor is to preserve it. This means that a project—such as a construction—cannot be denied based on religious reasons and if the applicant and the project are in line with extant laws:

> I am there, I would tend to say, the guarantor of secularism. Secularism is not about accepting religions but, on the contrary, about recognizing them. So, the first thing about the Mormons was to know if they were a cult or not. (Mayor)

He, therefore, consulted MIVILUDES with the conclusion that the Mormons were not a cult.

This is also attested by the spokespersons of the Mormons in France who also emphasize that MIVILUDES does not list them as a cult:

> We have a very good relationship with MIVILUDES. If you call them and ask, "What do you think of the Mormons?" there is no problem. (French Mormon spokesperson)

Without having further information about the relationship of the Mormon community and MIVILUDES, the narrative above indicates a certain openness of the Mormon community toward authorities and willingness to discuss.

On the other hand, these arguments are not shared by the president of the Association for the Defense of Families and Individuals Victims of

Cults in Yvelines (ADFI) who regrets the weak position of MIVILUDES and the "too close" links between this mission and the Mormons:

> MIVILUDES has nothing to say about the Mormons. That's it. It is an inter-ministerial mission. So, it has a fragile side. [...] It has a position that is also political because it is linked to the Prime Minister. So, it does not have the same freedom that our associations can have. (President cult defense)

Based on the testimonies of ex-Mormons that her association has collected, she finds

> [...] still a sectarian tendency. Yes, yes. It is a pyramidal, authoritarian organization and the control of the followers' thoughts is very strong. (President cult defense)

But she cannot prove a violation of the law either:

> [With] Mormons, it's much more subtle. There's tremendous control. There's tremendous pressure. There is no violation of the law. If people want to leave, they can leave. (President cult defense)

We can summarize, that officially and according to the French laws, Mormonism in France is perceived to be in line with the French legal norms, however, critics find this alignment borderline and remain suspicious. Therefore, Mormons can apply for any project if this project also respects the legal norms. According to the City of Chesnay, all the legal norms were respected and resulted therefore in the building permit of the temple. Furthermore, the mayor underlined that opposition was mostly of political nature and was finally not successful as any legal foundation of the opposition was missing.

Social Norms Influenced by the French Paradox of Secularism and Catholic Dominance

During our interviews, our interlocutors referred to social norms that are related to the perception of Mormons in France and that may have had an influence on the acceptance of or opposition to the construction of the temple in Chesnay.

The Mormon spokespersons mentioned that, in their opinion, the Catholic religion, on the one hand, and secularism, on the other hand, meant that Mormons were often perceived as a cult.

> What is specific to France is that Mormons are easily seen as a cult. Because in France, the Catholic religion was the state religion until recently. And in France, we are not used to religious plurality, that is to say, we accept the six main religions: the Jews, Protestants, Catholics, Muslims, Buddhists, Orthodox. But afterwards, everything else in France, these new religions, for them are cults and as in France secularism is very strong and there is sometimes an opposition to religion, an epidermal reaction, it is true that everything that was different was necessarily a sect. (French Mormon)

The mayor of Le Chesnay also suspects that an environment with a very Catholic and practicing population may have created opposition to the construction of the temple, especially as a protection of one's own religion.

> It was a difficult issue because the Mormons are in a very Catholic town, very religious. Le Chesnay is a very religious town. I myself am a catholic believer and a practitioner. And therefore, I had more opponents among the Catholics than atheists, let's be clear. (Mayor)

According to Mormon spokespersons, this caused a fear of invasion:

> The people in opposition (to the mayor) were relentless. And so, they spread the famous rumors, that is, I'm kidding but... "they're going to come with their handcarts. The 38,000 Mormons of France are coming. They're going to convert the whole city" and all that stuff. But people react to that. So, they are scared. (French Mormons)

The mayor of Le Chesnay assumes that the construction of any other religious monument other than Catholic would have provoked the same opposition:

> So, the arrival of the Mormons came up. I ask a question - if it had been a mosque, wouldn't it have caused so much noise? Of course, it would have caused as much noise. Let's be clear! It would have been a mosque we would have said the same! (Mayor)

Within the context of France and the specific location of the Mormon temple, there is evidence that a predominantly catholic society perceives Mormons as transgressors of catholic values and is, therefore, basically opposed to the manifestation of other religious values and artifacts. However, this is not shared by less practicing Catholics or atheists:

> I have no particular opinion on the Mormons. France is a free country. They can believe in whom or what they wish. Their liberty ends where my liberty or other people's liberty starts and vice versa. (French citizen 2)

Mormons themselves sometimes find themselves to defend certain Mormon values or correct Mormon myths which are different from traditional French social norms with regard to the status of women, technological interest, French gastronomical values. Some of differences in values are fact, some seem to be exaggerated or presented in a misleading way and some are untrue.

For instance, polygamy, although abolished in 1890, was cited during the construction of the temple and was put forward by a resident of Le Chesnay who was opposed to the construction:

> And you have a lady who stands up and says, "Me, I don't want a Mormon church. I don't want my husband to become a polygamist". (French Mormon)

The status of the woman in Mormonism is a point that is particularly advanced by ADFI, which refers to the testimonies of ex-Mormon women.

> I have had more interviews with female ex-Mormons than with men. And it's not surprising when you see the situation of the Mormon woman - it's still not funny. A woman is meant to serve her husband and have children. [...] Afterwards, mothers also take care of relief associations and so on - so good women's things. So, they take care of the girls' groups and all that. There are few women working. There may be more because you must earn a living. (President cult defense)

This narrative reflects the French social norm of gender equality and female emancipation which refer to historical events such as the emancipation movement around Simone de Beauvoir in the 1950s or the abortion law introduced by Simon Veil in 1975.

This position of the woman is explained differently by Mormons:

> Our doctrine is that God has divided the roles. He has given motherhood to women, and he has given the priesthood to men. But for us, the definition of priesthood is learning to serve. It's serving others, the priesthood. So, when you have that in mind, automatically we know that a woman serves naturally by her maternal instinct. [...] We were converted in 1977, here in the Paris region there were still many women who did not work. We had five children in France at that time. Now we have a few large families. But it's not a lot, and when you see our own married daughters, almost all of them are working and of course the number of children is decreasing, they have three children.... (French Mormon)

Thus, when it comes to the status of the woman, there are differences between Mormon communities and French society; however, it seems that the Mormon values have evolved over time in line with French values.

We asked the Mormon spokespersons and the two young missionaries we met at the temple if they felt discriminated against because of their faith. According to the two missionaries, respectively, from Toulon in France and North Carolina in the United States, neither of them felt discriminated against. They both attended a non-Mormon school and claim that their beliefs and principles such as abstinence from alcoholic beverages and chastity were tolerated and never discriminated against. This is also underlined by the narrative of a French Non-Mormon:

> I don't know a lot about them, but they seem to have some different values than the average French. But then vegan people also have different values when it comes to eating, very religious people have other values than less religious people and so on. I find differences more interesting than disturbing. (French citizen 1)

However, the spokespersons qualify this statement by saying that this is true for today, but that there was a time of discrimination in France, including losing one's job. According to them, young Mormons are now more daring to publicly affirm their belief, as are other denominations:

> There are times, there was a time up until the 1990s, when members didn't necessarily talk about their commitment to the Church at work. [...] if you're employed, they can't fire you for being Mormon. But there were on contract [...] and they lost their contract overnight, because in the

press a picture of them was published saying that he is the bishop of the Mormon parish and all the contracts stopped. But there are not thousands of cases like that. But there was until the 1990s, I think, a certain fear. That doesn't mean we were hiding. But it does mean that we didn't necessarily talk about our commitment. (French Mormon)

The Mormon spokespersons underlined that French Mormons remain French and that their political orientation, their educational level, the number of children reflect the French society.

In the end, a French Mormon remains French. He is French first and Mormon second. (French Mormon spokesperson)

The spokespersons also admit that people sometimes confuse Mormons with Amish and think that they give up a modern life, which is—according to them—not the case:

There are times the image of sad people, they are represented with black hats, because everything is confused with the Amish. So, they think we are Amish. And I'll have you know that I have this (he pulls out his cell phone) and when we have the scripture lesson on Sunday morning at church with the youth, the first thing we do is we take all the phones. Because if we don't, we can't do the lesson, you know... Like in school (he laughs). Just to say that we are not people of the Middle Ages. We are very technological. (French Mormon spokesperson)

During this qualitative study, about the controversial discussion about the construction of the Mormon temple in Le Chesnay, we were also interested in the reaction of the opponents of the temple, once it was built. According to the mayor of Le Chesnay, the opposition was rather political in nature, however, they pushed their campaign by suggesting that Mormons transgress precious French social values. The construction was a welcome affair by the opposition to undermine the mayor. The mayor stresses that he was elected for a new term despite the controversial discussion about the temple. He gives several reasons. On the one hand, he believes that the residents prefer the temple to a large social housing complex. On the other hand, he emphasizes the economic benefits for individuals who rent rooms to visiting Mormons. He, personally, emphasizes the pleasant contact he has had with Mormon leaders:

So, the Mormon project has already been able to respond to the urbanism and the environment of the neighborhood. But of all the residents next door, no one has complained. Today, everyone says thank you, it's very good [...] This makes the hotels and restaurants work, and the B&Bs. The Versailles people, they work a lot with the Mormons. They are very discreet. They were the first to complain and they are the first to profit. [...] I have met during the project Mormons of a very high quality. Afterwards, the philosophical and spiritual ideas of each person do not belong to everyone. (Mayor)

Mormon spokesmen say they have perceived a change in opposition to the construction of the temple thanks to a large information campaign on their part with neighborhood meetings and the possibility of visiting the temple before its inauguration:

In the end, they forced us to speak and, as early as 2012, to explain what the temple was, whereas we thought it was a subject not very easy to explain and we had said, 'We'll see when we get there.' And finally, from 2012–2013, we explained a lot of things, we showed a lot of things. And when the open house was held, a lot of people came, 47,000 visitors, including a large part of the surrounding area, including 2,000 V.I.P. People said that the building amazed them. Everyone came out saying: It is very beautiful; it is very clear. You can feel something calm. (French Mormon)

To summarize, the qualitative study demonstrated that the construction of the Mormon temple in Le Chesnay led to questioning by the Chesnay town hall, the neighboring population, and the associations regarding the compatibility of the temple and the Mormons with legal and social norms.

In France, Mormons do not officially transgress legal norms, although the ADFI advocates a sectarian tendency. Concerning social norms, we have noticed that certain characteristics of Mormon principles are not all in line with the values of French culture. However, it seems that non-Mormons follow the French philosophy of secularism where one's religion remains in the private sphere (Bertossi, 2011) resulting into tolerance and non-discrimination.

CONCLUSION

The construction of the Mormon temple in Le Chesnay, near Paris, sparked a controversial discussion about the Mormon community in France between 2012 and 2017. We, therefore, found it interesting to analyze how the French society, within a context of secularism, perceive the followers of Mormonism. We analyzed whether Mormons—according to French non-Mormons—act in (non-)conformity with social and legal norms in France. Our analyses show that Mormons do not violate French law. They do not transgress legal norms.

As far as social norms are concerned, Mormons do not live in isolation and in contradiction with French society. On the other hand, they defend certain values that are not followed with the same intensity by the French, such as abstinence from alcohol and tobacco and chastity before marriage. These differences seem to be clearly perceived but do not result, today, in discrimination of Mormons by the French. On the contrary, most of the people interviewed seem to accept Mormons despite these differences. However, a critical voice comes from the Association for the Defense of Families and Individuals Victims of Cults (ADFI), which regrets a strong control of individual thought among Mormons, especially regarding Mormon women, as well as a strong concentration of life on their community, leaving little time and space to build circles of friendship outside the Mormon community.

REFERENCES

Aleiss, A. (2010). Mormon influence, imagery run deep through twilight. *Huffington Post*. https://www.huffingtonpost.com/2010/06/24/mormon-influence-imagery_n_623487.html?guccounter=1

Atlantico. (2012). Twilight: Tout ce dont parle vraiment la saga (au-delà des vampires). https://www.atlantico.fr/decryptage/557402/twilight--tout-ce-dont-parle-vraiment-la-saga-au-dela-des-vampires-

Becker, H. S. (1963). *Outsiders: Studies in the sociology of deviance*. Free Press.

Bertossi, C. (2011). National models of integration in Europe: A comparative and critical analysis. *American Behavioral Scientist, 55*(12), 1561–1580.

Derocher, L. (2018). *Intervenir auprès des groupes sectaires ou des communautés fermées*. Presses de l'Université du Québec.

Durkheim, E. (1982). *The rules of sociological method*. Free Press.

Eglise de Jésus-Christ des Saints des Derniers Jours. (2011). *Brochure sur l'Eglise de Jésus-Christ des Saints des Derniers Jours*.

Europe1. (2017). *Qui sont les Mormons en France?* https://www.europe1.fr/soc iete/qui-sont-les-mormons-de-france-3291251

Flick, U. (2000). Episodic interviewing. In M. W. Bauer & G. D. Gaskell (Eds.), *Qualitative researching with text, image and sound* (pp. 75–92). Sage.

Gayral-Taminh, M. (2009). Une immigration invisible, gage d'intégration? *Ethnologie Française, 39,* 721–732.

Gillette, A. (1985). *Les Mormons—Théocrates du désert.* Editions Desclée de Brouwer.

Goosz, Y. (2017). Le débrief politique. Un Mormon dans l'équipe Fillon: "après tout c'est son droit, c'est la laïcité", selon ses soutiens. *Franceinfo.* https://www.francetvinfo.fr/replay-radio/le-debrief-politique/le-debrief-pol itique-un-mormon-dans-l-equipe-fillon-apres-tout-c-est-son-droit-c-est-la-lai cite-selon-ses-soutiens_2013212.html

Heath, E. A. (2011). The Gospel according to twilight. *Huffington Post.* https://www.huffingtonpost.com/elaine-a-heath/the-gospel-according-to-twilight_b_958630.html

Heckmann, F., & Schnapper, D. (2003). *The integration of immigrants in European societies: National differences and trends of convergence.* Lucius Verlag.

INED-INSEE. (2010). Trajectoires et origines: Enquête sur la diversité des populations de France. www.ined.fr/fichier/s_rubrique/19558/dt168_teo.fr. pdf

Jacobsson, K., & Löfmarck, E. (2008). A sociology of scandal and moral trans-gression: The Swedish 'Nannygate' Scandal. *Acta Sociologica, 51,* 204–2017. https://doi.org/10.1177/0001699308094166

Laffargue, O. (2013). Le christianisme, toujours première religion du monde. *BFMTV.* https://www.bfmtv.com/international/christianisme-tou jours-premiere-religion-monde-446572.html

L'Obs. (2017). On a visité le temple mormon du Chesnay: 4 questions sur un lieu secret. Publié le 08 avril 2017 à 11h50. https://www.nouvelobs.com/ societe/20170407.OBS7745/on-a-visite-le-temple-mormon-du-chesnay-4-questions-sur-un-lieu-secret.html

Mayer, J.-F. (1989). Jean-François Mayer, Nouveaux mouvements religieux: Une perspective historique et interculturelle. Dans S. Ferrari (Ed.), *Diritti dell'uomo e libertà dei gruppi religiosi* (pp. 17–40). CEDAM.

MIVILUDES. (2018). https://www.derives-sectes.gouv.fr/quest-ce-quune-d% C3%A9rive-sectaire/que-dit-la-loi

Mucchielli, L. (1999). La déviance, entre normes, transgression et stigmatisation. *Sciences Humaines, 99,* 20–25.

Passard, C., & Perl, P.-O. (2013). *Sciences Economiques et Sociales.* Bordas.

Rathgeber, T. (2013). *Ökumenischer Bericht zur Religionsfreiheit von Christen weltweit 2013.* Deutschen Bischofskonferenz. https://www.dbk.de/fil eadmin/redaktion/diverse_downloads/presse_2012/GT21_Oekum-Ber icht_web.pdf

Rigal-Cellard, B. (2000). Être français dans une Eglise d'origine américaine: Les Mormons de France. Dans C. Lerat, & B. et Rigal-Cellard (Eds.), *Les mutations transatlantiques des religions.* Les Presses de l'Université de Bordeaux.

Rigal-Cellard, B. (2003). Les cérémonies des mormons de nos jours: Mystère et initiation dans le temple. Dans M. Agostino, F. Cadilhon, P. Loupès (Eds.), *Fastes et cérémonies: L'expression de la vie religieuse, XVIe-XXe siècles* (pp. 163–186). Presses Universitaires de Bordeaux.

Smith, D. T. (2016). Predicting acceptance of mormons as Christians by religion and party identity. *Public Opinion Quarterly, 80*(3), 783–795.

Temple List. (2018). https://www.lds.org/temples/list?lang=eng

Vladescu, E. (2006). The assimilation of immigrant groups in myth or reality? Jean Monnet/Robert Schuman Paper Series 5(39). European Union Center.

Yang, I., & Bacouel-Jentjens, S. (2019). Identity construction in the workplace: Different reactions of ethnic minority groups to an organizational diversity policy in a French manufacturing company. *Organization, 26*(3), 410–431.

Yin, R. K. (2011). *Applications of case study research.* Sage.

INDEX